Navigation: Science and Technology

Volume 5

This series *Navigation: Science and Technology (NST)* presents new developments and advances in various aspects of navigation - from land navigation, marine navigation, aeronautic navigation to space navigation; and from basic theories, mechanisms, to modern techniques. It publishes monographs, edited volumes, lecture notes and professional books on topics relevant to navigation - quickly, up to date and with a high quality. A special focus of the series is the technologies of the Global Navigation Satellite Systems (GNSSs), as well as the latest progress made in the existing systems (GPS, BDS, Galileo, GLONASS, etc.). To help readers keep abreast of the latest advances in the field, the key topics in NST include but are not limited to:

- Satellite Navigation Signal Systems
- GNSS Navigation Applications
- Position Determination
- Navigational instrument
- Atomic Clock Technique and Time-Frequency System
- X-ray pulsar-based navigation and timing
- Test and Evaluation
- User Terminal Technology
- Navigation in Space
- New theories and technologies of navigation
- Policies and Standards

This book series is indexed in **SCOPUS** database.

More information about this series at http://www.springer.com/series/15704

Wei Zheng · Yidi Wang

X-ray Pulsar-based Navigation

Theory and Applications

Science Press
Beijing

Springer

Wei Zheng
College of Aerospace Science
and Engineering
National University of Defense Technology
Changsha, Hunan, China

Yidi Wang
College of Aerospace Science
and Engineering
National University of Defense Technology
Changsha, Hunan, China

The Funder information: National Natural Science Foundation of China (Grant No. 61703413).

ISSN 2522-0454 ISSN 2522-0462 (electronic)
Navigation: Science and Technology
ISBN 978-981-15-3295-5 ISBN 978-981-15-3293-1 (eBook)
https://doi.org/10.1007/978-981-15-3293-1

Jointly published with Science Press
The print edition is not for sale in China (Mainland). Customers from China (Mainland) please order the print book from: Science Press.
ISBN of the Co-Publisher's edition: 978-7-03-064771-9

This Springer imprint is published by the registered company Springer Nature Singapore Pte Ltd.
The registered company address is: 152 Beach Road, #21-01/04 Gateway East, Singapore 189721, Singapore

Foreword

Spacecraft increase in number with the development of space science and technology. If the position and velocity of those spacecraft are still provided by the ground-based tracking system, there are numerous human efforts and investment should be involved. In addition, the ground tracking is not capable of providing a timely response to some emergencies because of the distance between spacecraft and the Earth. Therefore, the autonomous navigation technique is important for future spacecraft. X-ray pulsar-based navigation is a promising spacecraft autonomous navigation method, which can make the spacecraft get rid of the support of the ground-based tracking system and the other artificial beacons. The NICER performed on the International Space Station, the Insight-Hard X-ray Modulation Telescope, and the X-ray pulsar-based navigation-01 satellite all has a common aim that is to verify the X-ray pulsar-based navigation. It might indicate a new surge of X-ray pulsar-based navigation.

The authors' research group has been working on X-ray pulsar-based navigation since 2004. They have proposed many methods to enhance the navigation performance of such navigation system. Now, the authors published their achievements, aiming to provide theoretical guidance and technical support for researchers working on the spacecraft autonomous navigation, especially X-ray pulsar-based navigation.

This book aims to investigate the X-ray pulsar-based navigation and to expand its application field. This book starts with introducing the X-ray pulsar-based spacecraft positioning/time-keeping/attitude determination methods, analyzes the error propagation mechanism and corresponding error compensation methods, proposes integrated navigation methods based on the information from X-ray pulsar and that from the other measurement sources, introduces the idea that autonomous navigation for spacecraft group with X-ray pulsar time difference of arrival, and finally designs a ground-based verification system. In addition, the methods in this book also provide a useful reference for solving related technical problems on spacecraft autonomous navigation.

I am glad to see this book get published and sincerely hope it will help upgrade the study on X-ray pulsar-based navigation.

Bao Weimin

May 2019

Bao Weimin
Academician of Chinese
Academy of Sciences
Beijing, China

Preface

Spacecraft navigation information nowadays is generally provided by the ground-based tracking and control system. However, the on-orbit spacecraft increase in number as space technologies develop, thus significantly burdening the ground-based tracking and control system and reducing the survivability of spacecraft in special cases. Therefore, greatly developing autonomous navigation technologies is in urgent demand in coping with increasingly complicated space missions and a key to enhance the survivability of spacecraft and to reduce operation costs.

X-ray pulsar-based navigation is a new spacecraft autonomous navigation method whereby the pulsar is performed as a celestial beacon to provide the reference information for spacecraft to estimate its state including position, attitude, and time. Compared with satellite navigation, X-ray pulsar-based navigation is not just applicable to near-earth space but also insensitive to manual inference. Compared with the traditional celestial navigation method, X-ray pulsar-based navigation can simultaneously provide complete navigation information such as the position, attitude, and time. X-ray pulsar-based navigation shows obvious characteristics in the X band and can avoid jamming from various signals in space. Compared with radio pulsars, X-ray pulsar with high energy at the X band can guarantee the minimization of detectors that have sufficient flux sensitivity and temporal-spatial resolution.

The X-ray pulsar-based autonomous navigation method was proposed initially in 1980s. From the end of the twentieth century to the beginning of the twenty-first century, a relatively complete navigation framework has been gradually developed. The Flight Dynamics and Control Team in the National University of Defense Technology has been studying the theory and methodology of X-ray pulsar-based navigation since 2004 and is one of the first teams in China that researches X-ray pulsar-based navigation. This book introduces our research achievements over the past 13 years, combining with the latest global development in the field.

This book is divided into seven chapters. Chapter 1 first provides a schematic picture on the autonomous navigation for spacecraft and reviews the development of X-ray pulsar-based navigation. Chapter 2 provides the basic knowledge supporting X-ray pulsar-based navigation; Chap. 3 analyzes how to process X-ray pulsar signals when the spacecraft is stationary and when the spacecraft is orbiting; Chap. 4 investigates the error within X-ray pulsar-based navigation and proposes methods to overcome them; Chap. 5 introduces the integrated navigation by fusing the X-ray pulsar signal and other information sources; Chap. 6 employs the X-ray pulsar time difference of arrival to fulfill the spacecraft autonomous navigation; Chap. 7 designs and constructs an X-ray pulsar-based navigation ground simulation and verification system.

We hereby would like to sincerely thank the National Natural Science Foundation of China, National Science and Technology Major Project, National High Technology Research and Development Program of China (863 Program), and relevant experts in China for their supporting research presented in this book.

We referred much literature of scholars home and abroad and cited such literature to the best of our knowledge when writing this book. We hereby want to express our sincere appreciation to them.

We would also like to thank the academicians Bao Weimin and Wei Ziqing for their consistent care and support. We would especially thank Academician Bao Weimin for contributing Foreword of this book. We thank researchers Zhang Shuangnan, Shuai Ping, Liu Siwei and Lu Fangjun, Prof. Li Xiaoping and Prof. Fei Baojun, and Associate Professor Yao Guozheng for their help in specific research work and their precious suggestions on this book. We thank Mr. Qian Jun, the editor, for his great efforts in getting this book published.

X-ray pulsar-based navigation is an interdisciplinary research field involving navigation theories, orbital dynamics, X-ray astronomy, high-energy physics, microelectronics, etc. It is developing continuously in its theories and applications. Authors welcome any corrections from peers and readers for mistakes in this book.

Changsha, China
October 2017

Wei Zheng
Yidi Wang

Contents

Chapter 1
Introduction

1.1 Basic Concept of Spacecraft Autonomous Navigation System

1.1.1 Definition of Spacecraft Autonomous Navigation System

A navigation system for spacecraft can be viewed as an autonomous one if it is self-contained, could provide real-time operation, does not radiate, and is independent of the ground support. In practice, the definition of autonomous navigation system for spacecraft can be loosen to be that a navigation system which gets rid of the support from the ground-based system is the autonomous navigation system.

1.1.2 Necessity of Autonomous Navigation Systems

The current spacecraft are commonly tracked by the ground-based systems. However, as aerospace missions become increasingly elaborate, the future spacecraft will be encouraged to experience unknown environment. In this case, spacecraft are expected to autonomously cope with emergencies, since the ground-based systems might not provide timely emergence responses due to the huge distance between spacecraft and the Earth. The autonomous navigation technique is a key technique to fulfill the autonomous operation [1].

1.1.2.1 Necessity for Earth-Orbiting Spacecraft

Earth-orbiting spacecrafts, which include various satellites, spaceships, space maneuvering spacecraft, and orbit transfer vehicles, perform important roles in the modern

W. Zheng and Y. Wang, *X-ray Pulsar-based Navigation*, Navigation: Science and Technology 5, https://doi.org/10.1007/978-981-15-3293-1_1

society. With the development of aerospace science and techniques, the Earth-orbiting spacecraft launched in recent years increase sharply in type and number, burdening ground systems.

In order to reduce the manual effort involved in the ground-based support and to enhance the survivability when a spacecraft facing hostile environment, the spacecraft is suggested to have the autonomous navigation capability. For the spacecraft in the near-Earth space with orbital altitude of lower than 3000 km, the autonomous navigation can be well implemented by the global navigation satellite system (GNSS). However, for those spacecraft orbiting on the high Earth orbit, the received signal of GNSS is instable, and is difficult to be employed to fulfill the function of autonomous navigation. Nevertheless, high-Earth-orbit spacecraft, with an advantage in the orbit altitude, play a role in daily life more important than low-Earth-orbit spacecraft. How high-Earth-orbit spacecraft could have autonomously positioning performance is a hot issue for the study on spacecraft autonomous navigation.

1.1.2.2 Necessity for Deep Space Explorers

Since the beginning of the 21st century, the countries in the world all attached importance to the deep space exploration, which help human beings probe the secrets of the universe. As the flight distance and duration of mission are huge, deep space explorers put new demands on each key techniques supporting the missions, especially on the navigation technique [2].

At present, most of the international deep space exploration missions are supported by the ground-based radio tracking systems, among which the Deep Space Network (DSN) of the United States (US) is the most famous one. The radio tracking systems determine the position of deep space explorers by measuring the distance and radial velocity of the deep space explorer relative to the ground-based system [3].

However, this method has inevitable shortcomings as follows:

(1) It involves frequent communications between the ground-based system and the deep space explorer. There would be a huge time delay within the communication caused by the huge distance between the ground-based system and the deep space explorer. The time delay within the communication from the Mars to the Earth can be up to 45 min and that of communication concerning exploration for the Jupiter, Saturn and other celestial bodies would be larger. In addition, communication signals might be blocked or be disturbed by celestial bodies providing radio radiation. In this case, the ground-based system cannot provide a timely response to the request of deep space explorers facing emergence.
(2) It involves numerous manual efforts. An increasing number of deep space explorers would aggravate the burden of ground-based systems. A deep space exploration mission usually continues for several years or even several decades, during which numerous manual efforts and investment are needed by the ground-based systems to ensure the success of the mission.

(3) It cannot provide an accurate and real-time navigation result. For deep space missions such as approaching, flying over and impacting target celestial bodies, the position, velocity, and attitude of deep space explorers should be accurately known. However, those target celestial bodies are far away from the Earth, and the measurement accuracy of ground-based tracking technique would reduce by 4 km if the distance between the spacecraft and the Earth grows an astronomical unit (AU). Thus, the above missions are difficult to be accomplished if only the support of ground-based systems is available.

In order to extend the field of deep space exploration and to ensure the survivability of deep space explorers which might lose the contact with the ground-based systems, it is benefit to develop the autonomous navigation technique.

1.1.2.3 Necessity for Navigation Satellite Constellations

For objects on the ground or in the air, the navigation satellites are the most common approach to provide them with a high-precision position and a reference time. The ephemeris errors and clock errors of navigation satellites are the important factor that affects the positioning results based on observing navigation satellites. The current ephemeris and parameters of clock error of navigation satellites are corrected and uploaded by the ground-based systems. If there is any problems occurred in the ground-based systems, the performance of navigation satellite constellations would degrade. Hence, it is promising to allocate the function of autonomous navigation to the navigation satellite constellations.

For navigation satellite constellations, the autonomous orbit determination can be well approached by measuring the distance between satellites provided by inter-satellite links [4]. However, the measurement of relative distance can only accurately determine the relative positions of satellites but cannot resist the constellation as a whole to rotate in the inertial system [5].

Thus, it is significant to develop an autonomous navigation system for navigation satellite constellations to resist the whole rotation in the inertial system.

1.2 Three Main Types of Spacecraft Autonomous Navigation Systems

1.2.1 Inertial Navigation System

The inertial navigation system (INS) could provide the position and attitude information of spacecraft by measuring the apparent acceleration and the rotation speed of body system with respect to the inertial system and by propagating the dynamics model [6]. Featured by being independent of external information, free from interference, and elusive, INS has been widely utilized in aerospace, marine and military fields.

INS can be classified into gimbaled inertial navigation systems (GINS) and strapdown inertial navigation systems (SINS), according to the way how to gain the inertial measurements. An INS usually consists of a gyro and an accelerometer. There are two types of gyro that are widely applied, including mechanical gyro and optic gyro. The mechanical gyro senses the angular speed or angular displacement via the directing property and procession of mechanical rotors [7]. In 1852, Foucault, a French physicist, preliminarily proposed the concept of gyros. Since the 20th century started, mechanical gyro technique was stimulated by the growing requests of military and industrial applications and developed rapidly, bringing out the buoyancy gyros and electrostatical gyros [8], among which buoyancy gyros included liquid floated gyros, gas-floated gyros, magnetically suspended gyros, etc. [9]. To reduce the manufacture cost, vibration gyros were invented [10]. Vibration gyros include hemispherical resonator gyro, quartz tuning fork vibrating gyro, micro electro-mechanical system (MEMS) gyro, and etc. [11]. Optic gyro are classified into three categories: laser gyro, fiber-optic gyro and integrated optic gyro [12], among which laser gyro and fiber-optic gyro have been widely used while integrated optic gyro is still in the phase of development, but is promising in application. Besides mechanical gyro and optic gyro, the cold atom interferometry gyro benefited from the rapid development of optics technique is developing rapidly and is of great prospect [13].

The impact of error within the gyro on the positioning performance of INS is a cubic function of time. In order to improve the navigation performance of INS, it is necessary to understand profoundly the INS error model and to compensate the impact of errors besides improving the hardware manufacture quality [14]. Inertial navigation error models will change in parameters as the service environment, which will reduce the reliability of parameters calibrated in laboratories. In this case, some methods such as missile-borne tests, rocket sled tests and vehicle tests can be utilized to systematically verify the inertial navigation error model [15, 16].

1.2.2 Celestial Navigation System

Celestial navigation system (CNS) could determines the position and attitude of a spacecraft by measuring the position or direction information related to celestial bodies. This system is featured by strong autonomy, immunity from interference, high reliability, and shows an advantage because its navigation error does not accumulate over time [17].

CNS was first applied to the navigation for ships, and was introduced into the field of spacecraft navigation in 1950s, being benefited by the rapid development of electronic, computer and space techniques. Both the Apollo program and the space station of the Soviet Union utilized CNS.

The sole utilization of directional information of stars could only determine the attitude of a spacecraft. In order to determine the position of the spacecraft, the directional information of the spacecraft with respect to a nearby celestial body has to be obtained. For Earth-orbiting satellites, the nearby celestial body is commonly selected as the Earth or the moon. There are two ways to gain the directional information of the satellite with respect to the selected nearby celestial body, i.e., direct horizon sensing and indirect horizon sensing [17].

The direct horizon-sensing navigation method provides the position and attitude of spacecraft by employing horizon sensors and star sensors. In 1960s, the United States Air Force developed the first satellite autonomous navigation program (283 program) [18], which comprised three strapdown gyros, a star sensor, and a horizon sensor. The positioning performance of this system was mainly limited by the undesirable accuracy of the horizon sensor, and could only achieve an accuracy of 2 km. In 1973, the United States Air Force launched the Space Sextant-autonomous Navigation and Attitude Reference System (SS/ANARS) [19] consisting of two optical telescopes installed on a 3 degree of freedom rotation platform, between which one telescope was for tracking the bright edge of the moon and the other one was for tracking a known star. This system was designed to have an attitude-determination accuracy of $0.6''$ and a positioning accuracy of 224 m.

The indirect horizon-sensing navigation method, which is based the stellar refraction, could provide the position of a spacecraft by employing high-precision star sensors. In 1979, the US developed the Multi-mission Attitude Determination and Autonomous Navigation (MADAN) [20] which could provide real-time and continuous inertial attitude and orbit information through three star sensors and was featured by full autonomy. The target positioning accuracy of the system was 0.9 km for low Earth orbit satellites and 9 km for high Earth orbit satellites.

Besides the direct and indirect horizon-sensing methods, the Microcosm, Inc., in 1989, developed a system which determined the real-time orbit and attitude of a spacecraft by measuring the Earth, Moon and Sun with satellite-borne special autonomous navigation sensors—Microcosm Autonomous Navigation System (MANS) [21]. MANS could provide an autonomous navigation service for medium-low Earth orbit satellites. Its navigation sensor was an improved version of a conical scanning infrared earth sensor, featured by light weight, low power consumption, low cost, and etc. In March, 1994, this system was loaded by the Space Test Platform-0 spacecraft to verify its feasibility and key techniques involved. Unfortunately, the data of MANS had to be transmitted to the ground-based system for further analysis due to the failure in the satellite-borne computer. The positioning accuracy of MANS was estimated to be about 200–500 m.

When the 21st century began, the US, France and Japan initialed a new upsurge in deep space exploration. The CNS for deep space explorers is gradually becoming a critical backup navigation system supporting the ground-based system in deep space exploration missions. Both the "Deep Impact" mission and "Hayabusa (MUSES-C)" detector utilized CNS to enhance the autonomous survivability of deep space explorers [22, 23].

1.2.3 Navigation Satellite System

Navigation satellite system can be viewed as a combination of CNS and radio tracking technique. It determines the position and velocity of a spacecraft by measuring the distance and Doppler frequency from the spacecraft to navigation satellites, the position and velocity information of which can be previously gained by the public ephemerides of navigation satellites. As being featured by low cost and high-accuracy, navigation satellite systems have been widely applied to the current aerospace missions.

In 1958, the US developed a navigation satellite system named Transit, also known as Navy Navigation Satellite System (NNSS), which used navigation satellites instead of ground-based stations as the navigation reference beacons. This system came into service in 1964, was used for public use in 1967 [24], but was phased out since 1990, which was caused by the successful applications of the Global Positioning System (GPS).

GPS is a second-generation global navigation satellite system that was first proposed in 1973 and was put into the full operation in 1994 [25]. This system overcomes the shortcomings of the first-generation global satellite navigation system, such as the failure to perform continuously positioning, low accuracy and long interval for positioning, and can provide the high-precision position, velocity, reference time, and attitude of clients. GPS comprises 24 operational satellites and 3 backup satellites, at orbits with inclination of 55° and average height of 20,200 km and period of 11 h 58 min. Navigation satellites are distributed on 6 equally-spaced orbits, with 4 satellites on each orbit. The right ascensions of ascending nodes between two orbits differ by 60° and the arguments of perigee of the satellites on the adjacent orbits differ by 30° (Fig. 1.1).

Besides GPS, the Soviet Union declared to construct GLONASS (Global Navigation Satellite System) and launched the first satellite on October 12th 1982 (Fig. 1.2) [26]. After the Soviet Union collapsed, Russia took this system over and continued to develop it. In 1995, Russia declared the initial operation of GLONASS. The whole GLONASS constellation is also comprised of 24 operational satellites and 3 backup satellites, but unlike GPS, these satellites, spaced with each other by 120°, are distributed on 3 orbits with inclination of 64.8°, height of 19,100 km and period of 11 h 15 min. 8 satellites are on each orbit and distributed with interval of 45°.

The European Unit (EU) announced to develop GALILEO satellite navigation system in 1999 (Fig. 1.3) [27]. The GALILEO satellite constellation comprises 30 navigation satellites (including 3 backup satellites) which are evenly distributed on 3 orbits with inclination of 56° and height of 23,616 km. 9 operational satellites and 1 backup satellite are on each orbit with period of 14 h 4 min.

In China, Beidou-1 navigation satellite system project was officially approved in 1994, two satellites were successfully launched respectively on Oct. 31st and Dec. 21st, 2000 and a backup satellite was also successfully launched on May 25th, 2003 (Fig. 1.4). Beidou-1 consists of 3 geosynchronous satellites, among which two operational navigation satellites are respectively positioned at 80° E and 140° E

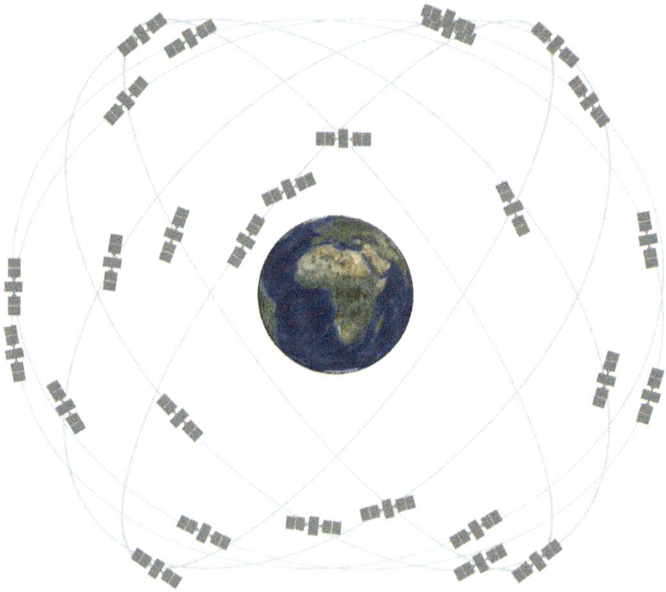

Fig. 1.1 GPS Constellation (from www.gps.gov)

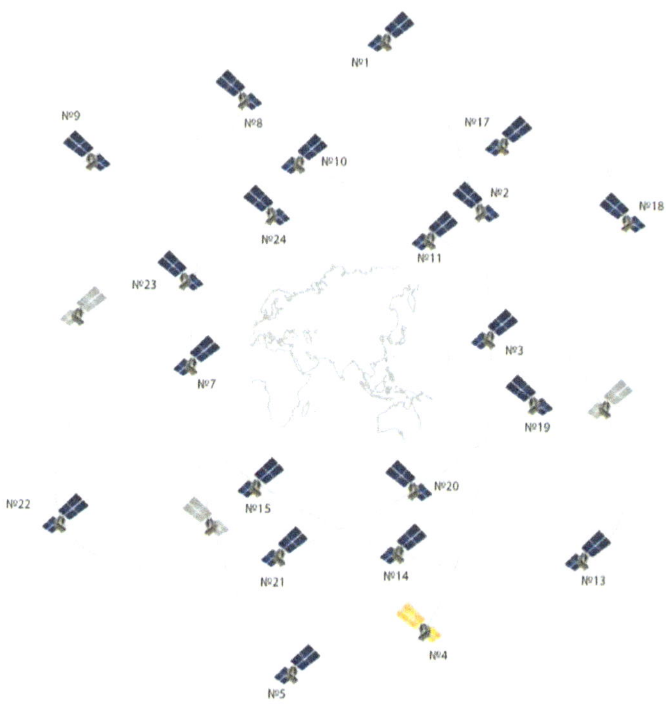

Fig. 1.2 GLONASS constellation (from http://gssc.esa.int)

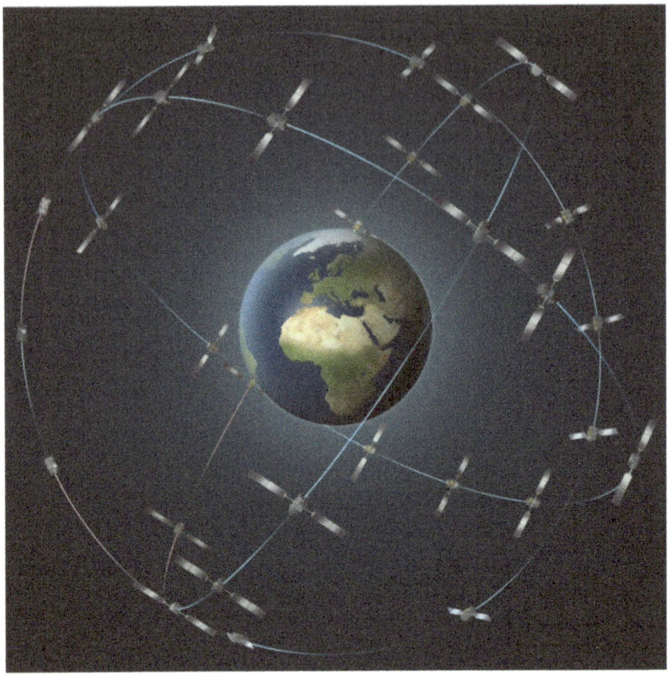

Fig. 1.3 GALILEO navigation constellation

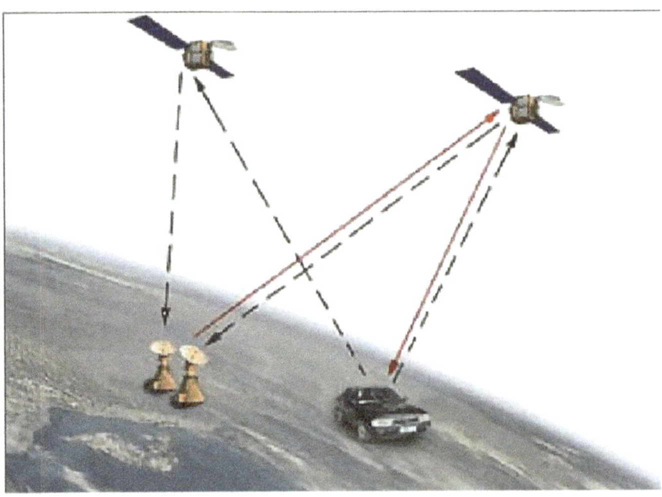

Fig. 1.4 Schematic diagram of navigation performed by "Beidou-1" navigation satellite system (*Image credit* sinodefence)

while the backup satellite is positioned at 110.5° E. The Beidou-1 Navigation Satellite System is an active system. For positioning, a client should first send a request to the ground station which will compute the position of client and then it to the client. It causes the client has to carry a receiver including the function of transmitter, which would increase the weight, cost and power consumption.

Hence, Beidou-2 Navigation Satellite System started to be constructed in 2004 and the construction was planned to be performed in two steps: first, establishing the local Beidou Navigation Satellite System in 2012 to form a local covering ability and offer function of passive positioning, navigation and timing services to China and the Asian-Pacific region; second, expanding the local Beidou Navigation Satellite System from 2013 and completely establishing the Beidou Navigation Satellite System around 2020 which can offer passive positioning, navigation and timing services to the whole world [28]. On December 27th, 2012, the first step of the Beidou-2 Navigation Satellite System was completed and the system has the functions as designed.

The final version of Beidou Navigation Satellite System will consist of 5 geostationary satellites and 30 non-geostationary satellites, with main functions of passive positioning, velocity measurement, single/double way timing service and short-message communications. This system offers positioning, velocity-measuring and timing services with the accuracy higher than 10 m, 0.2 m/s and 20 ns respectively, and performs short message communication with a capacity of more than 120 Chinese characters per time.

1.3 Review of X-Ray Pulsar-Based Navigation

1.3.1 Brief Introduction of Pulsar

As a branch of neutron star, pulsars are the products of supernova explosions caused by massive stars at the end of their lifetime. For a pulsar, the spinning axis and the magnetic axis do not coincide and its two magnetic poles simultaneously emit electromagnetic radiation beams as shown in Fig. 1.5. When a spacecraft is swept over by the beams, it could receive a pulsed signal like a ship receives signals from a lighthouse. Pulsars are spinning at periodicities with excellent long-term stabilities, and some millisecond pulsars could even match the current atomic clocks. Most of time, a pulsar could simultaneously radiate at different wavebands, such as optical, radio, X-ray, and γ-ray. The X-ray radiation is recommended to facilitate navigation, as it needs sensors quite smaller than devices that function in radio and optical wavelengths [29]. In pulsar astronomy, pulsars that could provide X-ray and radio radiation are usually called X-ray and radio pulsars, respectively.

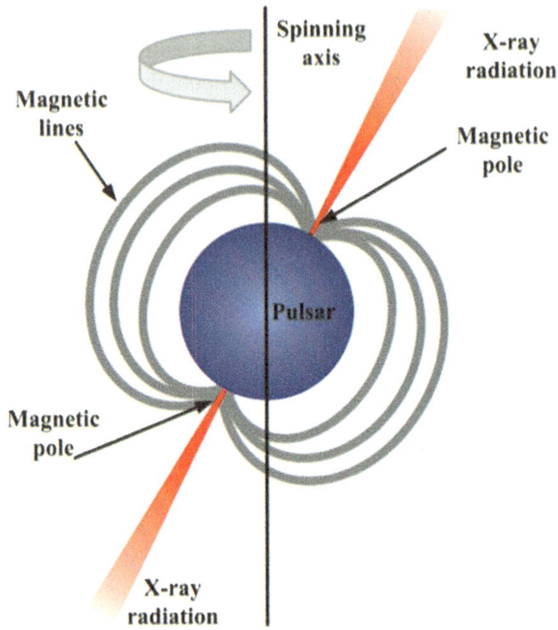

Fig. 1.5 Lighthouse model of pulsar

1.3.2 Brief Introduction of X-Ray Pulsar-Based Navigation

X-ray pulsar-based navigation (XPNAV) is a developing and promising autonomous navigation technique. XPNAV works by utilizing the X-ray radiation of pulsars. Compared with the current celestial navigation systems, which work mainly by measuring the position or direction information of celestial bodies, XPNAV utilizes the timing information of pulsars and thus its navigation performance is little affected by the distance between the pulsar and the spacecraft. Compared with navigation satellite system, XPNAV is applicable to the near-Earth and deep space, and resists the artificial inference.

By employing the timing information of pulsars, XPNAV could accomplish three types of applications distinct from the other autonomous navigation systems:

- Autonomous maintenance of time reference
  The onboard atomic clock is expected to provide the accurate timing information for the whole spacecraft. However, the unavoidable frequency drift of onboard atomic clock might degrade all the operations of spacecraft. In this case, pulsars could be employed as natural time references to reduce the impact of onboard clock error by means of pulsar timescale and pulsar-aided time-keeping. Pulsar timescale is a timescale utilizing the high periodicity of pulsar. If pulsars are observed for 10 years, the stability of ensemble pulsar timescale could achieve the order of 10^{-14} which was comparable to the estimated long-term stabilities of the

best atomic clocks. Pulsar-aided time-keeping is realized by steering the onboard atomic clock to the observed pulsar.

- Autonomous navigation for navigation satellite constellation
  As illustrated in Sect. 1.1.2.3, the autonomous navigation for satellite constellation is currently implemented by using inter-satellite link. However, inter-satellite links could only well determine the relative positions of satellites within a constellation but cannot resist the rotation of the whole constellation. In this case, pulsars can be viewed as natural "anchors" which could provide absolute direction reference for the whole satellite constellation in the inertial coordinate system. Compared with the previous ground-based "anchor" method, pulsar-based method could completely get rid of human interference.
- Accurate autonomous interplanetary navigation
  The current autonomous navigation method available for deep space explorers (DSEs) in the Sun-centered cruise is accomplished by measuring the stellar angular distance with an accuracy of on the order of arc-minute which corresponds to a positioning accuracy of above a thousand kilometer. In contrast, XPNAV works by handling the timing information of pulsar. If the timing information has an accuracy of higher than 0.1 ms, which is easy to be achieved, the positioning accuracy of XPNAV is higher than 30 km. Thus, XPNAV could achieve an accurate autonomous navigation for DSEs performing the Sun-centered cruise.

In order to realize XPNAV, three key techniques, i.e., navigation pulsar database, the technique of pulsar signal detection and processing, and the technique of navigation theory, are needed.

1.3.3 Famous Programs on XPNAV

These are some famous programs have been performed to study and verify XPNAV.

1.3.3.1 Programs of the US

(1) **Unconventional stellar aspect (USA) experiment**

The US experiment, also called NRL-801 [30], aimed at verifying the feasibility of determining the orbit and attitude of a spacecraft via space X-ray sources and the feasibility of keeping the time reference via X-ray pulsars. It was jointly developed by the Stanford Linear Accelerator Center (SLAC) and the United States Naval Research Laboratory (NRL), and was launched as a payload of the Advanced Research and Global Observation Satellite (ARGOS) in 1999. Although the experiment was designed to last for 3 years, it had to be aborted in the November 2000, as the onboard X-ray sensor broke down.

(2) **XNAV program**

X-ray source-based navigation for autonomous position determination (XNAV) was initiated in 2004. It planned to develop an autonomous positioning, attitude determination and timekeeping services with the use of space X-ray sources, and aimed at providing a reliable backup for GPS and an effective navigation means for spacecraft in the near-Earth and deep space.

The whole program consisted of three phases [31]: Phase I (Concept Feasibility) that aimed at characterizing pulsar, designing the prototype X-ray sensor and position/attitude algorithm; Phase II (Development) that aimed at developing pulsar characterization, payload processing, and performing flight demonstration on the International Space Station (ISS) or an aerospace plane; and Phase III (Mission Operations) that aimed at deploying the technique developed by XNAV. Unfortunately, XNAV was unfunded out of some non-announced causes at the end of Phase I.

(3) **XTIM program**

The Defense Advanced Research Projects Agency (DARPA) proposed the X-ray Timing (XTIM) program, which was implemented by the Lockheed Martin corporation, in 2010. XTIM aimed at constructing a global pulsar time system and performing tests on a geosynchronous orbit. XTIM could distribute the positioning and timekeeping information to client spacecraft in order to help them determine the position as well as the reference time.

(4) **SEXTANT program**

The Station Explorer for X-ray Timing And Navigation Technology (SEXTANT) is a technology-demonstration enhancement to the Neutron-star Interior Composition Explorer (NICER) mission [32]. SEXTANT is funded by the space technology mission directorate game changing development program office of National Aeronautics and Space Administration (NASA). NICER is an X-ray astrophysics mission of opportunity to the ISS that will undertake a fundamental investigation of extremes in gravity, material density, and electromagnetic fields of rapidly spinning neutron stars via time-resolved X-ray spectroscopy. SEXTANT seeks to demonstrate real-time onboard XPNAV, with a stretch objective of achieving 1 km orbit determination accuracy along any direction using up to 4 weeks of observations. SEXTANT has been successfully launched to the ISS on the June 3rd, 2017. In the January 2018, NASA announced the SEXTANT has demonstrated the XPNAV onboard.

1.3.3.2 Programs of the Europe

Based on the study achievement of X-ray observation and astrophysics, the European scientists studied the feasibility of XPNAV and developed different types of X-ray sensors.

1.3.3.3 Programs of China

China launched the first dedicated X-ray navigation satellite (XPNAV-01) in November 2016. XPNAV-01 successfully observed the pulsar PSR B0531+21 and the received data has been published in May 2017. Meanwhile, Chinese astronomers have verified XPNAV via the data collected from POLAR loaded on TianGong-2.

1.3.4 Progresses of Key Techniques

1.3.4.1 Progress of Navigation Pulsar Database

Pulsars, signal sources of XPNAV, should be carefully modeled and characterized. For pulsars applied to XPNAV, the position, period evolution parameters, and X-ray radiation properties are necessary to be calibrated and contained in the navigation pulsar database.

As the radio radiation could penetrate the dense atmosphere of the Earth, ground-based radio large telescopes have been developed to determine the position and period evolutions of pulsars by investigating the radio radiation. The information related to the X-ray radiation, such as the profile, has to be determined by launching X-ray satellite, since X-rays could be absorbed by the atmosphere.

(1) **Observation on radio pulsars**

The radio astronomy started when Reber successfully built the first radio telescope in 1937. The radar technology that rapidly developed over the World War II indirectly promoted the radio telescope technology. Since the first pulsar was discovered, many countries started to observe radio pulsars through single antenna large radio telescopes. Table 1.1 lists the eight famous radio telescopes. Now, those single antenna radio telescopes combine with the other new developed telescopes to form some pulsar timing observation arrays, such as the Parkes pulsar timing array (PPTA), the American nanohertz observatory for gravitational waves (NanoGrav), and the European pulsar timing array (EPTA). All the above arrays are combined to be the international pulsar timing array (IPTA). Observation through the Arecibo Telescope led to the discovery of the first binary pulsar, the first millisecond pulsar, and the first pulsar planets. The Parkes telescope discovered more than half of the currently known pulsars.

To increase the time resolution, the scientists in the United Kingdom (UK) successfully developed the multi-element radio-linked interferometer network (MERLIN) in 1980. MERLIN uses the long baseline multi-antenna interferometry to increase the aperture of a radio telescope, and finally it is equivalent to an single radio telescope with an aperture of 217 km. Motived by MERLIN, the very long baseline interferometer (VLBI) was proposed. Unlike MERLIN, the radio telescopes employed by a VLBI are not linked together, but are required to observe the same radio source

Table 1.1 Main radio telescopes for radio pulsar observation

PTA	PPTA	NanoGrav		EPTA				
Telescope	Parkes telescope	Arecibo telescope	Green bank telescope	Effelsberg telescope	Lovell telescope	Sardinia radio telescope	Westerbork synthesis radio telescope	Nançay radio telescope
Country	Australia	US	US	Germany	UK	Italy	Holland	France
Aperture [m]	64	305	100	100	76	64	96	94

independently and simultaneously. When an observation is finished, the outputs of all telescopes are sent to a correlator to produce the final result. Since 1980s, European VLBI Network and the very long baseline array (VLBA) have been established in succession. At present, VLBI is gradually becoming a main means of radio pulsar observation.

(2) **Observation on X-ray pulsars**

The discovery of the first extra-solar X-ray source, Scorpius X-1 (Sco X-1), marked the start of X-ray astronomy. There are many X-ray satellites that have been launched since then. We briefly introduce some remarkable satellites.

In 1973, the first X-ray satellite named Uhuru was launched by US. It completed the first comprehensive and uniform all-sky X-ray survey, catalogued 339 X-ray sources, and discovered the diffuse X-ray emission from clusters of galaxies.

The second high energy astronomy observatory satellite (HEAO-2) launched in 1978 was the first fully imaging X-ray telescope. It performed the first high resolution spectroscopy and morphological studies of supernova remnants, and the first medium and deep X-ray surveys.

The European space agency's (ESA's) X-ray observatory (EXOSAT) was launched in 1983 [31]. It studied the variability of X-ray and discovered quasi-periodic oscillation (QPO) in low-mass X-ray binary (LMXRB) and X-ray pulsars.

In 1990, Germany, UK and US jointly developed the Roentgen Satellite (ROSAT) for X-ray surveys. It accomplished the first X-ray all-sky survey at the soft X-ray band by employing an imaging telescope with a sensitivity of about 1000 times better than Uhuru. It cataloged more than 150000 objects and discovered the X-ray from comets and the pulsations from Geminga.

The Rossi X-ray Timing Explorer (RXTE) was launched on the December 30th, 1995 [32]. It has the highest time resolution (up to 1 μs) and high sensitivity among the explorer satellites of NASA so far. The combination of the loaded Proportional Counter Array (PCA) with a large effective area (6250 cm^2) and moderate energy resolution makes it possible for RXTE to study the timing variability of X-ray sources. RXTE discovered the spin periods in the LMXRB, the X-ray afterglows from Gamma Ray Bursts, and initialized a new era for X-ray timing variability research.

To refine the structures and energy spectra of X-ray sources, In Jul. and Dec., 1999, NASA and ESA respectively launched an X-ray imaging satellite: Chandra and XMM-Newton. The large effective areas make them applicable to study on the faint X-ray sources and greatly improve the observation on X-ray sources.

In July 2017, China launched its own hard X-ray astronomy satellite, Hard X-ray Modulation Telescope (HXMT), which will draw the first highly-accurate hard X-ray sky map. Through joint efforts of the world, people will definitely understand the physical mechanism of X-ray pulsars more deeply, thus providing basic theoretical instructions and data support for the future XPNAV applications.

(3) **Catalog of navigation pulsars**

In order to facilitate XPNAV, researchers proposed various criteria to select X-ray navigation pulsars, and corresponding built several pulsar databases for navigation.

There are two types of pulsars that are selected by all the databases: (1) the rotation-powered pulsars, and (2) the pulsars that simultaneously radiate at the radio and X-ray wavelengths. It is because that the measurement model for rotation-powered pulsar is much simpler and more applicable than binaries and that the simultaneous radiation guarantees the position and period of pulsar be measured by the ground-based telescopes.

The current famous databases of navigation pulsars are database of Microcosm, Inc (8 pulsars), database of ESA (10 pulsars), database of Sheikh (25 pulsars) and database of SEXTANT (11 pulsars). Since many pulsars are included in multiple databases at the same time, we delete the repeated ones, and provide the final navigation pulsars database as the union of the above 4 databases. Table 1.2 lists the parameters of selected pulsars the distribution of which in the equatorial coordinate system are shown in Fig. 1.6.

1.3.4.2 Process of Pulsar Signal Detection and Processing

In order to illustrate the navigation principle of XPNAV, the pulsar signal was assumed to be continuous and pulsed. However, practical pulsar signals are so weak that a spacecraft could merely record a series of photon TOAs. In addition, there are various celestial bodies that radiate X-rays and might interfere the reception of pulsar signal. Thus, it is necessary to efficiently collect the photon TOAs and accurately extract a pulse TOA out of the recorded photon TOAs.

(1) **X-ray sensor**

As shown in Fig. 1.7, the basic elements of an X-ray sensor are the optics, which collects the radiation entering the sensor, and the detector, which transfers the collected photons to be electrons with arrival times recorded by an atomic clock. And then, the photon TOAs are obtained.

1) **Optics**

The common methods using in X-ray optics are focusing optics and collimator.

- Focusing optics

  X-rays tend to be absorbed in the most materials without altering direction much, since the refractive indices of all materials are very close to 1 for X-rays. However, it is possible for X-rays to perform efficient total external reflection from a highly-polished boundary if the X-rays reach the surface at a small grazing incidence angle. Compared to collimator, the focusing optics could achieve a higher resolution imaging and a higher telescope sensitivity since X-rays are focused on a small area. The current optical geometry to facilitate the X-ray focusing optics are the Wolter type and the Lobster eye type.

  In 1952, Wolter designed three structures that employ two reflections to focus X-rays, among which the type I (or Wolter-I) is commonly employed.

Table 1.2 Navigation pulsars

Name	Galactic Longitude [°]	Galactic Altitude [°]	Period [s]	Flux [ph/cm^2 s]
J1939 + 2134	57.51	−0.29	0.00156	4.99×10^{-5}
J1959 + 2048	59.20	−4.70	0.00160	8.31×10^{-5}
J0218 + 4232	139.51	−17.53	0.00232	6.65×10^{-5}
J1824 − 2452	7.80	−5.58	0.00305	1.93×10^{-4}
J0751 + 1807	202.73	21.09	0.00347	6.63×10^{-6}
J0030 + 0451	113.14	−57.61	0.00487	1.96×10^{-5}
J2124 − 3358	10.93	−45.44	0.00493	1.28×10^{-5}
J1012 + 5307	160.35	50.86	0.00525	1.93×10^{-6}
J0437 − 4715	253.39	−41.96	0.00575	6.65×10^{-5}
J0537 − 6910	279.55	−31.76	0.01611	7.93×10^{-5}
J0534 + 2200	184.56	−5.78	0.03340	1.54
J1959 + 2408	68.77	2.82	0.03953	3.15×10^{-4}
J1302 − 6350	304.18	−0.99	0.04776	5.10×10^{-4}
J0539 − 6944	279.72	−31.52	0.05037	5.15×10^{-3}
J1811 − 1926	11.18	−0.35	0.06467	1.90×10^{-3}
J0205 + 6449	130.72	3.08	0.06568	2.32×10^{-3}
J1420 − 6048	313.54	0.23	0.06818	7.26×10^{-4}
J1617 − 5055	332.50	−0.28	0.06934	1.37×10^{-3}
J0835 − 4510	263.55	−2.79	0.08929	1.59×10^{-3}
J1826 − 1433	18.00	−0.69	0.10145	2.63×10^{-3}
J1709 − 4428	343.10	−2.68	0.10245	1.59×10^{-4}
J1124 − 5916	292.04	1.75	0.13531	1.70×10^{-3}
J1930 + 1852	54.10	0.27	0.13686	2.16×10^{-4}
J1513 − 5908	320.32	−1.16	0.15023	1.62×10^{-2}
J1846 − 0258	29.71	−0.24	0.32482	6.03×10^{-3}
J1808 − 3658	355.39	−8.15	0.00249	0.329
J1932 + 1059	47.38	−3.88	0.2265	1.4×10^{-4}
J0633 + 1746	195.13	4.27	0.23709	2.05×10^{-3}
J0659 + 1414	201.11	8.26	0.38487	6.9×10^{-3}
J1057 − 5226	285.98	6.65	0.14315	4.68×10^{-3}
J1024 − 0719	251.7	40.52	0.00516	1.59×10^{-3}

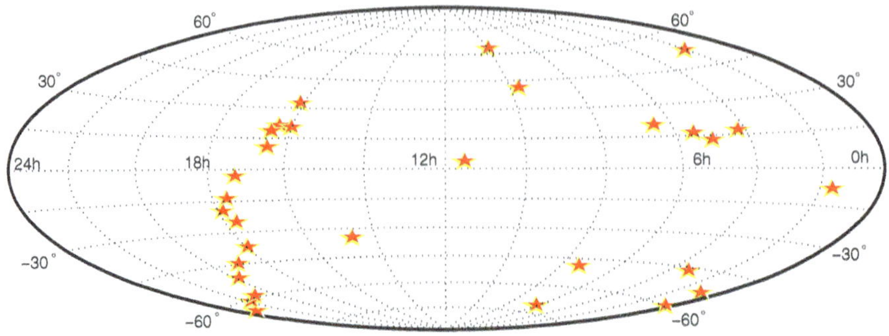

Fig. 1.6 Distribution of navigation pulsars in the equatorial coordinate system

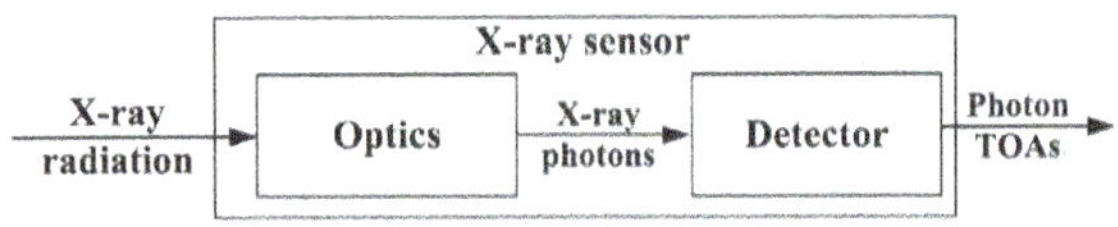

Fig. 1.7 Schematic working flow of X-ray sensor

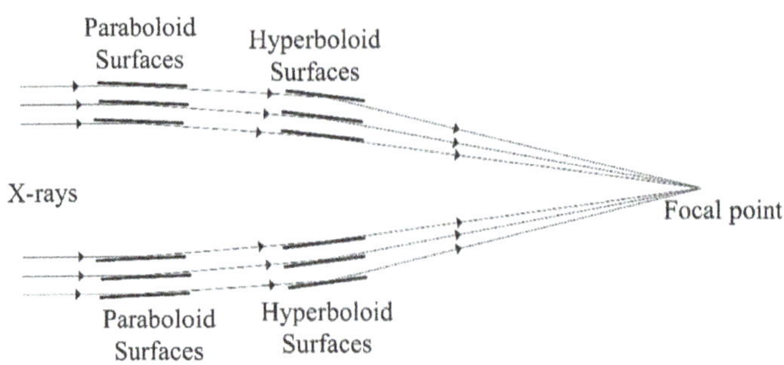

Fig. 1.8 Structure of Wolter-I

Figure 1.8 three-layers structure of Wolter-I. Wolter-I is typically composed of parabolic and hyperbolic shells. X-rays are first reflected by the convex sides of parabolic shells and then are focused to a focal point by a second reflection at the convex sides of hyperbolic shells. However, limited by the refractive indices of materials for X-rays, the field of view (FOV) of Wolter-I is usually less than 1°. Wolter-I has been widely applied to various X-ray satellites, such as HEAO-2, EXOSAT, Chandra, and XMM-Newton.

The Lobster-eye optical system originates from the bionics study on the compound eyes of lobsters or the other crustaceans. It is composed of multi-channel grazing incident mirrors (Fig. 1.9). Being benefitted from the symmetrical sphere structure, the Lobster-eye optical system has a FOV of up to 180°. The common structures of Lobster-eye optical systems are Schmidt type and

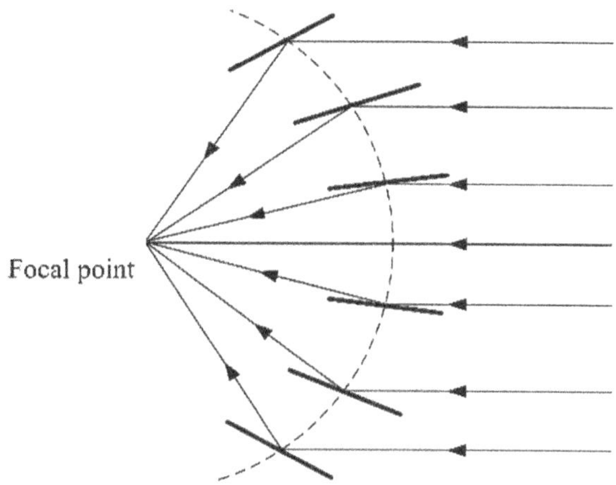

Fig. 1.9 Structure of Lobster-eye system

Angel type. The Schmidt type consists of two orthogonal stacks of reflectors. The Angel type consists of numerous tiny square cells located on the sphere and is similar to the reflective eyes of lobsters. The Angel optical system has a complicated structure but has a focal distance shorter than Schmidt type and a lighter weight. The Astronomical Institute Ondrejov manufactured two Lobster-eye models and performed ground-based experiments to testify the excellent performance of Lobster-eye optical system in 2004. NASA revealed its study on the Lobster-eye optical system, ISS-Lobster, in 2012. ISS-Lobster was still in the conceptual study phase and was intended to simultaneously provide the wide FOV, high position resolution, and high sensitivity.

- Collimator
  An X-ray collimator is an array of parallel channels, the FOV of which can be defined by the ratio of the length and the width. The collimator is positioned above an underlying detector the area of which is identical to that of collimator. Figure 1.10 shows the conceptual structure of collimator.

2) **Detector**
   An X-ray detector could indirectly detect X-ray photons by recording electrons which are the products of the photoelectric effect and the Compton effect caused by the interaction between X-ray photons and certain materials in the detector [33]. The common types of X-ray detectors are Proportional Counter (PC), Charge-Coupled Device (CCD), Swept Charge Device (SCD), Silicon Drift Detector (SDD), Micro-Channel Plate (MCP), and Scintillator [34].

- PC
  A PC works by measuring the particles of ionizing radiation which are produced when X-rays enter the PC and ionize the gas within it [35]. The energy of the particles produced by the ionization of gas is proportional to that of X-ray photons, and this is the very reason for its name. PC is featured by simple structure and easy processing, which make PC be widely employed by

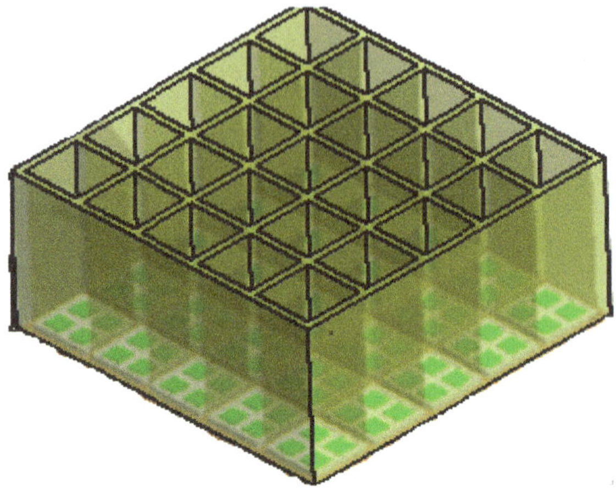

Fig. 1.10 Conceptual structure of collimator

the earlier X-ray satellites, such as Uhuru, HEAO-1, EXOSAT, ROSAT, and ARGOS. However, it has a low detection efficiency and poor energy resolution, and needs a high voltage to ionize gas. In addition, the detection performance would degrade if there is gas leakage.

- CCD
  CCD is a semiconductor detector. CCD is featured by low power consumption, being capable of miniaturization, and providing photon position resolution better than 100 μm. However, it needs to work in a low temperature and could only provide a time resolution of about 0.01 s. Given that CCD records the 2-dimensional position of a photon, it takes almost 1 s to readout, which is unacceptable by a practical navigation mission. CCD has been applied to Chandra, XMM-Newton, and Suzaku.
- SCD
  SCD is a special CCD-type X-ray detector [36], which could shorten the readout time to be less than 1 ms by discarding the 2-dimensional position information of photons but only keeping the time information. SCD is featured by the high detection efficiency, good energy resolution, and low power consumption. But, it needs a complicated circuit to support a large sensitive area. SCD has be applied to HXMT.
- SDD
  SDD is also a semiconductor detector. SDD employs a series of ring electrodes that causes charge carriers to drift to a small collection electrode. SDD is featured by high count rate and high energy resolution. SDD has been successfully applied to the Curiosity [37].
- MCP
  MCP is a photoelectric multiplier. It could work in the room temperature, and provides a time resolution of on the order of nanosecond. However, it has to be supported by a high-voltage power source, and does not have an energy resolution. MCP has been applied to ROSAT, Chandra, and HXMT.

- Scintillator
  When charged particles or photons enter a scintillator, atoms in the scintillator will be ionized and excited, emitting a fluorescence. The scintillator detects X-ray photons or charged particles by measuring the fluorescence. Scintillator has been applied to Suzaku, Vela-5B, OSO-7, OSO-8, HEAO-1, CGRO and RXTE [38].

(2) **Pulsar signal processing**

The pulsar signal is so weak that a spacecraft could merely record a series of photon TOAs instead of a continuous pulsed signal. The current study on pulsar signal processing performs mainly on period search and extraction of pulse TOA out of the recorded photon TOAs.

First of all, the period search is to find the period of a pulsar signal which is quite important for recovering the pulsar profile. Given that the photon TOAs are unevenly recorded, the classical Fourier transform cannot be straightforwardly employed to find the period. There is a long history in period search algorithms for the unevenly sampled problem, such as the Lomb-Scargle periodogram, string length method, auto-correlation method, phase dispersion minimization, analysis of variance, and information-theoretical methods. For a detailed review on period search algorithm, please refer to.

When the spacecraft is assumed to be stationary or perform a uniform linear motion, there are two types of pulse TOA calculation methods: epoch folding type method as well as the direct use of photon TOAs. The process of epoch folding type method proceeds by three steps: (1) recovering an empirical profile via epoch folding; (2) estimating the initial phase via methods such as cross-correlation method, adaptive filter, nonlinear least square algorithm, and fast near-maximum likelihood estimator; and (3) obtaining the pulse TOA as a ratio of the estimated initial phase to the period of pulsar signal. If the period of pulsar signal is not accurate, it must be found out by the above period search methods. The direct use of photon TOAs type method performs by maximizing the log-likelihood function derived based on the stochastic properties of photon TOAs.

For orbiting spacecraft, Golshan et al. proposed a phase tracking algorithm to estimate the pulse TOA and time-varying frequency. This method approximates the orbit of spacecraft to be a piece-wise and constant model to guarantee the frequency within each piece could be well approximated to be constant. And then, the pulse TOA and frequency at each piece can be readily estimated by well-developed methods. Finally, a digital phase locked loop (DPLL) is employed to track the frequency varying over pieces. The DPLL can be further modified to be two-dimensional Kalman filter to cope with the noise within DPLL. The phase tracking algorithm works well for young pulsars such as PSR B0531+21 but fails when applied to faint pulsars. In order to validate faint pulsars for XPNAV, we modified the propagation model of pulsar signal to be a term of the position of spacecraft, and utilized the orbital dynamics information of spacecraft to linearize the propagation model. Compared with the previous phase tracking algorithm, the method does not approximate the orbit of spacecraft to be piece-wise and constant model which hinders the utilization of faint pulsar.

1.3.4.3 Process of Navigation Theory

Until now, the development of navigation theory of XPNAV experiences three phases:

(1) Phase of concept introduction (1971–1980). Reichley et al. first proposed the idea that the radio pulsars can be employed for navigation. In 1974, Downs proposed an interplanetary navigation method using radio pulsar, and found that the spacecraft could have a positioning accuracy of around 150 km. However, in this case, a spacecraft has to load an radio telescope with a diameter of at least 25 m and keeps observing pulsars for 24 h. In addition, it is difficult to extract the radio signal of pulsar out from the radio radiation from the other celestial sources. Regarding that the X-ray radiation could guarantee the miniature of navigation sensor, Chester and Butman first introduced the concept of spacecraft navigation using X-ray pulsar in 1981.
(2) Phase of preliminary formation (1980–2005). In 1993, Wood suggested to establish an X-ray navigation system that can provide the position, attitude and time reference for spacecraft and that encourages the USA experiment. In 2004, ESA analyzed the basic principle of XPNAV and pulsar signal models, and investigated the feasibility of engineering implementation of XPNAV. Sheikh made great contributions to XPNAV. He widely surveyed the technologies related to XPNAV, developed an navigation pulsar database, and put forward a practical measurement model for XPNAV with the consideration of general relativistic effect, Roemer effect and Shapiro effect. The achievement of Sheikh built an almost complete framework of XPNAV.
(3) Phase of refinement (2005-present). The study on XPNAV turned from the pure theoretical study to the application implementation. Sheikh et al. analyzed the positioning performance of combing the XPNAV and deep space network, revealing that DSEs could have positioning accuracies of less than 3 km in this case. The US and China both revealed their flight demonstration of XPNAV, i.e. SEXTANT and XPNAV-01. In 2015, the Max-Planck-Institute for Extraterrestrial Physics hosted a conference on XPNAV. The researchers from US, ESA, China etc. discussed the unsolved problems as well as the potential solutions.

References

1. Hu X (2002) Autonomous navigation theory and application. National University of Defense Technology Press, Changsha
2. Wiliam HB (2005) Deep impact mission design. Space Sci Rev 117(1–2):23–42
3. Mancuso S (2004) Vision based GNC systems for planetary exploration. ESA ESTEC, Netherlands

4. Cai Z, Han C, Chen J (2008) Constellation rotation error analysis and control in long-term autonomous orbit determination for navigation satellites. J Astronaut 29(2):522–528
5. Zhang Y (2005) Study on autonomous navigation of constellation using inter-satellite measurement. National University of Defense Technology, Changsha
6. Barbour N (2001) Inertial components-past, present, and future. AIAA Guidance, Navigation and Control Conference. American Institute of Aeronautics and Astronautics, Montreal, Canada
7. Fan L (2009) Research on autonomous navigation method for automated transfer vehicle. National University of Defense Technology, Changsha
8. Deng S, Xiao Z, Yan L, Huang A (2012) The status and prospects of integrated optical gyroscopes and related topics. Physics 41:179–185
9. Wrigley W (1977) History of inertial navigation. J Inst Navig 24(1)
10. Su Z, Li Q, Li K (2010) Inertial technology. National Defense Industry Press, Bei Jing
11. Barbour N. MEMS for navigation—a survey, plenary session. ION National Technical Meeting, Long Beach
12. Usui R, Ohno A. Recent progress of fiber optic gyroscopes and applications at JAE. Optical fiber sensors conference technical digest, USA, pp 11–14
13. Zou P, Yan S, Lin C, Wang G, Wei C (2013) Research status and prospects of cold atom interferometry gyroscope in inertial navigation fields. Mod Navig 8:263–269
14. Wen Y (2012) Research on the navigation method of high orbit automatic transfer spacecraft. National University of Defense Technology, Changsha
15. Wang H (2004) The study of algorithm and simulation for ring laser gyro strapdown inertial navigation system for rocket. National University of Defense Technology, Changsha
16. Wen Y (2007) Research on vehicle test technique in validation of INS error model. National University of Defense Technology, Changsha
17. Jiancheng F, Xiaolin N (2006) Principles and applications of astronomical navigation. Beijing University of Aeronautics and Astronautics Press, Bei Jing
18. Chory MA, Hoffman DP, Lemay JL (1986) Satellite autonomous navigation status and history. IEEE Position Location and Navigation Symposium, Las Vegas
19. Lowrie JW (1979) Autonomous navigation system technology assessment. AIAA 17th Aerospace Science Meeting
20. Chory MA, Hoffman DP, Major CS (1986) Autonomous navigation-where we are in 1984. AIAA guidance and control conference
21. Li J, Chen Y (1997) Advances in the spacecraft autonomous navigation technology. Aerosp Control 15(2):76–81
22. Riedel JE, Bhashkaran S, Synnott SP, et al. Navigation for the new millennium: autonomous navigation for deep space 1. Deep space 1 technology validation report—autonomous optical navigation, pp 48–65
23. Kawaguchi J, Hashinmoto T, Kubota T et al (1997) Autonomous optical guidance and navigation strategy around a small body. J Guid Control Dyn 20(5):1210–220
24. Xi X, Zhang S, Wang H (2003) Developments in satellite navigation systems. J Spacecr TT&C Technol 22(1):1–6
25. Wang J (2006) Review and prospect on the development of GPS/GNSS. Geotech Investig Surv 3:55–60
26. Liao C, Ge B, Liu H (1999) Russia global navigation satellite system, overview, status and development trend. Satell Appl 4:60–64
27. Flament P. GALILEO: a new dimension in international satellite navigation. 46th international symposium electronics, Zadar, Craotia, pp 6–14
28. Tang J (2013) Developing and applying analysis of BeiDou navigation satellites regional system. GNSS World of China 38(5):47–52
29. Zhu C (2003) Astronomy course. Higher Education Press, Beijing
30. Wood KS (1993) Navigation studies utilizing the NRL-801 experiment and the ARGOS satellite. In: International society of optical engineering proceedings
31. Pine DJ. X-ray navigation for autonomous position determination. 14 Feb 2013. http://sites.nationalacademies.org/DEPS/ASEB/DEPS_061644

32. NICER/SEXTANT team named goddard chief technologist's innovators of the year. 16 Feb 2013. http://www.usra.edu/news/pr/2011/nicer/
33. Shuai P, Li M, Chen S, Huang Z (2009) Principle and method of X-ray pulsar-based navigation. China Astronautic Publishing House, Beijing
34. Hu H (2011) X-ray photon counting detector for X-ray pulsar-based navigation. Graduate University of the Chinese Academy of Sciences, Beijing
35. Charpak G, Sauli F (1984) High-resolution electronic particle detector. Ann Rev Nucl Sci 34:285–350
36. Lowe BG, Holland AD, Hutchinson IB et al (2001) The swept charge device, a novel CCD-based EDX detector: first results. Nucl Instrum Methods Phys Res A 458:568–579
37. Liu L (2014) Orientation parameters determination of the navigation satellites constellation via X-ray pulsars measurement. National University of Defense Technology, Chang Sha
38. Da Rocha J, Lanceros-Mendez S (2006) 3-D modeling of scintillator-based X-ray detector. Sens J 6(5):1236–1242

Chapter 2
Fundamential of the X-Ray Pulsar-Based Navigation

2.1 Space-Time Reference Frame

2.1.1 Coordinate System

2.1.1.1 Conventional Celestial Sphere Reference System

A reference system is needed to represent the locations of celestial bodies and to establish ephemerides. The reference system includes a set of theories and data processing methods, a set of models and constants, and a corresponding reference frame.

In 1991, IAU decided to use the International Celestial Reference System (ICRS) in place of the Fundamental Katalog No. 5 (FK5) reference system which was built for measurement of stars, with mean equator plane and mean equinox of J2000.0 as the start point of declination and right ascension, respectively. As the position of the mean equinox has to be calculated by using the well determined positions of celestial bodies, FK5 reference system has large direction errors in coordinates. In contrast, the coordinate directions of ICRS are defined by the positions of extragalactic radio sources at J2000.0 (JD2451545.0, 12 h on January 1st, 2000) determined by VLBI, and the OXY plane of ICRS approaches J2000.0 mean equator plane. Compared to the stars employed to build FK5 reference system, extragalactic radio sources comprising ICRS have almost no proper motion, therefore, ICRS can be regarded as a geometrical frame which is not time-varying.

IAU2006 Resolution B2 introduces the Barycenter Celestial Reference System (BCRS) and Geo-centered Celestial Reference System (GCRS), in which the former's origin is located in the solar system barycenter, in the same direction as ICRS, while the latter's origin in the geocenter, in the same direction as BCRS.

IAU decided that ICRS should be realized by the International Celestial Reference Frame, (ICRF) and maintained by the International Earth Rotation and Reference System Service (IERS). In 1998, IERS proposed ICRF1, which comprised 608 radio sources in which 212 sources were used to define the coordinate directions of ICRS.

W. Zheng and Y. Wang, *X-ray Pulsar-based Navigation*, Navigation: Science and Technology 5, https://doi.org/10.1007/978-981-15-3293-1_2

The position accuracy of those definition sources is about 0.25 mas, and the direction of coordinates determined by these radio sources is more accurate than 0.02 mas. In 2009, IAU decided to use ICRF2 in place of ICRF1 since 2010. ICRF2 includes 3414 radio sources with positional accuracy of app. 40 μas and coordinates direction stability of app. 10 μas.

IAU2000 Resolution B1.2 recommended Hipparcos Catalogue to be used as the major realization of ICRS over the optical band, and named it the Hipparcos Celestial Reference Frame (HCRF).

Besides radio sources and stars, the ephemerides of the major planets and moon in the solar system (e.g. DE series ephemeris of JPL) can be considered as a dynamic realization of ICRS. Reference frames comprising such ephemerides are referred to be the dynamic reference frames which cannot self-maintain for a long time without adding new observation data. This is the very shortcoming of dynamic reference frames.

Compared to the standard ICRS, the dynamic frame has direction deviation. Thus, the ephemeris information provided by DE series ephemeris unavoidably contains errors with respect to ICRS. Additionally, since DE series ephemeris is employed in pulsar timing, these ephemeris errors will lead to pulsar parameter errors. Therefore, it is extremely essential to take into account the influence of pulsar-based navigation in the research of reference system timing the pulsar. Chapter 5 will elaborate on the effect of ephemeris errors over pulsar-based navigation by approximating the difference between different DE ephemerides to the ephemeris errors of celestial bodies.

2.1.1.2 J2000.0 Reference System

J2000.0 reference system has its origin located in the solar system barycenter or the geo-center, with J2000.0 mean equator plane as the reference plane, and X axis pointing to J2000.0 mean equinox [1]. The J2000.0 reference system with origin in the geo-center is also referred to as J2000.0 Earth Centered Inertial Coordinate System J2000.0 (ECI J2000.0). The J2000.0 reference system is also known as the FK5 reference system, because its direction is determined by FK5 star catalogue. Although IAU recommended ICRS to take place of FK5 reference system, the J2000.0 reference system is still commonly used in current applications and studies related to spacecraft dynamics and navigation.

To guarantee the continuity of reference system, the coordinates of ICRS should be as close as possible to those of FK5 reference system in J2000.0 epoch. In order to connect ICRS to KF5, the radio source 3C 273B, which is employed to build ICRS, is assumed to be of right ascension 12 h 29 m 6.6997 s in the FK5 reference system, then the relationship between ICRS and FK5 reference system can be determined by the deviation of celestial pole and zero point deviation of longitude.

2.1.2 *General Relativistic Time System*

In the framework of Newtonian mechanics, space and time are separate, and several time systems such as sidereal time, solar time and ephemeris time have been defined according to the observation on the motion of the Sun or the Earth [2]. In contrast, X-ray pulsars locate far away from the solar system, and the whole solar system is a Riemannian space where the path of light is distorted as a result of the presence of giant celestial bodies in the solar system. Thus, the impact of general relativity on the time system must be taken into consideration. In the framework of the general theory of relativity, space and time are integrated and the definitions and transformation between time systems become complicated accordingly.

2.1.2.1 Proper Time and Coordinate Time

The general relativistic space-time theory defines two concepts: proper time and coordinate time. Proper time, which has real physical meanings, is measured by an ideal standard clock in the local space-time where the observer is. Coordinate time is a "timelike" variable, which is obtained by calculating the time metric in the global space-time reference system and can be used as global time coordinates, is an abstractive concept. For spacecraft autonomous navigation using X-ray pulsars, the position information of a spacecraft is determined by comparing the difference in the time when the same pulse signal arrives at the spacecraft and at the solar system barycenter. According to the general theory of relativity, the times shall be compared in the framework of coordinate time.

In the general theory of relativity, the coordinate system is defined by metric tensors related to the gravitational field and matter distribution. The four-dimensional invariant element of length ds^2 and metric tensors in the space-time are correlated by

$$ds^2 = g_{\mu\nu} dx^\mu dx^\nu \tag{2.1}$$

where the metric tensor $g_{\mu\nu}$ is the function of x^μ. The relationship between the gravitational field and the space-time metric can be determined by the Einstein field equations

$$G_{\mu\nu} \equiv R_{\mu\nu} - \frac{1}{2} g_{\mu\nu} R = 8\pi G T_{\mu\nu} \tag{2.2}$$

where R is the scalar curvature, $R_{\mu\nu}$ is the Ricci tensor, $G_{\mu\nu}$ is the Einstein tensor and $T_{\mu\nu}$ is the energy-momentum tensor of matter.

Due to complexity of matter and energy distributions and high-order nonlinearity of the Einstein field equations, the strict solution of the field equations cannot be reached. Therefore, the space-time metric used for actual application needs to be approximated. In 1991, in the Resolution A4 of the 21st General Assembly, IAU recommended for the first time that the general theory of relativity being used as the theoretical basis for the space-time reference system [1]. In 2000, IAU proposed complete post-Newtonian space-time metrics and their applicable space-time transformation models at the 24th General Assembly [3].

For the static weak gravitational field in the solar system, terms lower than c^{-2} are omitted and the following space-time metrics can be applied [4]

$$\begin{cases} g_{00} = 1 + \frac{2U}{c^2} \\ g_{0i} = 0 \\ g_{ij} = \delta_{ij}\left(-1 + \frac{2U}{c^2}\right) \end{cases} \tag{2.3}$$

where U is the sum of Newtonian potentials in the gravitational field of interest. It is a negative value and becomes 0 at infinity. δ_{ij} is the Kronecker symbol (when $i = j$, $\delta_{ij} = 1$; when $i \neq j$, $\delta_{ij} = 0$).

The space-time metric can define the relationship between proper time and coordinate time. Suppose that the space-time coordinates of a spacecraft in a certain global coordinate system are (t, x^i), its space-time metric coefficient is $g_{\mu\nu}$, its velocity in the coordinate system is v^i and the proper time on the spacecraft is τ.

According to Eq. (2.1), we have

$$\begin{aligned} ds^2 &= c^2 d\tau^2 = g_{\mu\nu} dx^\mu dx^\nu \\ &= (g_{00} + 2g_{0i}v^i/c + g_{ij}v^i v^j/c^2)c^2 dt^2. \end{aligned} \tag{2.4}$$

Thus,

$$d\tau = \sqrt{g_{00} + 2g_{0i}v^i/c + g_{ij}v^i v^j/c^2} dt. \tag{2.5}$$

This is the theoretical relationship between coordinate time and proper time, which obviously depends on the selected space-time metrics. Hence, there is not only one space-time metric or coordinate time.

Substituting Eq. (2.3) into Eq. (2.5), the integral transformation from proper time to coordinate time can be described as

$$\frac{d\tau}{dt} = 1 + \frac{U}{c^2} - \frac{1}{2}\left(\frac{v}{c}\right)^2. \tag{2.6}$$

2.1.2.2 Definition of International Atomic Time

With the development of technology, the definitions of the time measurement and time scale are improved continuously. In 1955, the appearance of the atomic clock realized a great improvement on the time measurement. In 1958, Markowitz et al. defined the length of a second in the atomic time according to the ephemeris time. In 1967, Conference Generale des Poids et Mesures (CGPM) strictly defined the length of a second in the atomic time and used it as an SI unit which is still applicable now.

In 1971, the atomic time scale kept by the International Time Bureau was used as the time base and named as International Atomic Time (TAI). TAI started at 0 h 0.0039 s on January 1, 1958 (UT1). In 1980, the Consultative Committee for the Definition of the Second (CCDS) gave the relativistic definition of TAI.

Suppose

$$W = U - v^2/2 \tag{2.7}$$

is the gravitational potential of an object in the earth-centered coordinate system, where $v = \omega_e R$ is the ground rotation speed (R is the geocentric distance).

According to the definition given by the CCDS, TAI is the coordinate time on the geoid and its second is the SI second. Hence, supposing TAI is t', according to Eq. (2.6), we have

$$\frac{d\tau}{dt'} = 1 + \frac{U - U_0}{c^2} - \frac{1}{2}\frac{v^2 - v_0^2}{c^2} = 1 + \frac{W - W_0}{c^2} \tag{2.8}$$

where W_0 is the gravitational potential of the object on the geoid.

Due to the instability of actual atomic clocks, TAI is kept by using atomic clock times in multiple countries worldwide through a certain algorithm. With Eq. (2.8), atomic clock times in those countries can be converted to the time on the geoid. TAI, as a very important time scale, is the specific realization of relativistic time and serves as the bridge between non-relativistic time and relativistic time.

2.1.2.3 Terrestrial Time (TT) and Geocentric Coordinate Time (TCG)

(1) **Terrestrial time (TT)**

IAU changed Terrestrial Dynamical Time (TDT) into Terrestrial Time (TT) in 1991. TT is a conceptual time scale obtained by using an ideal clock on the geoid and defined specifically as follows: the reference time base of the apparent geocentric ephemeris is TT; TT differs from Geocentric Coordinate Time (TCG) by a constant rate and its measuring unit is the same as the SI second (Le Système International d'Unités) on the geoid; for continuity with its predecessor ephemeris time, TT is 0 h 00 m 32 s.184 on January 1, 1977 the moment when TAI is 0 h 00 m 00 s on January 1, 1977.

(2) **Geocentric coordinate time (TCG)**

Geocentric Coordinate Time (TCG), as the coordinate measurement scale for Geocentric Celestial Reference System (GCRS), corresponds to Terrestrial Time (TT) at 0 o'clock on January 1, 1977. Based on Eq. (2.6), the transformation between TCG and TT is

$$TCG = TT + L_G(JD - 2443144.5) \times 86400\,\text{s} \tag{2.9}$$

where $L_G = 6.969290134 \times 10^{-10}$ is the scale factor between TCG and TT, and JD is the Julian day for time transformation (the same below).

2.1.2.4 Barycentric Coordinate Time (TCB) and Barycentric Dynamical Time (TDB)

(1) **Barycentric Coordinate Time (TCB)**

Barycentric Coordinate Time (TCB) is introduced to describe the motion of celestial bodies in the solar system (with the barycenter of the solar system as the origin) in the framework of non-rotation relativity. TCB and TCG match TAI when TAI is 0 o'clock on January 1, 1977. Based on Eq. (2.6), the transformation between TCB and TCG is

$$TCB = TCG + \frac{L_C(TT - TT_0) + P(TT) - P(TT_0)}{1 - L_B} + c^{-2}\boldsymbol{v}_E(\boldsymbol{x} - \boldsymbol{x}_E) \tag{2.10}$$

where $\boldsymbol{x}_E$ and $\boldsymbol{v}_E$ are the position vector and velocity vector of the geocenter in the barycenter coordinate system; $\boldsymbol{x}$ is the position vector of the spacecraft in the barycenter coordinate system; $L_C = 1.48082686741 \times 10^{-8}$ and $L_B = L_C + L_G = 1.55051976772 \times 10^{-8}$; TT_0 corresponds to TAI when TAI is 0 o'clock on January 1, 1977. The time ephemeris TE405 can be used for obtaining the value of $P(TT) - P(TT_0)$ [5]. Equation (2.10) is only applicable to the near-earth space.

(2) **Barycentric dynamical time (TDB)**

TDB (Barycentric Dynamical Time) only differs from TT in periodic items. The formula for transformation between TCB and TDB is

$$TCB = TDB + L_B(JD - 2443144.5) \times 86400\ \text{s} + P_0 \tag{2.11}$$

where $P_0 \approx 6.55 \times 10^{-5}$ s and $L_B = L_C + L_G = 1.55051976772 \times 10^{-8}$.

The length of a second in TT is as the same as that in TDB, so the time scale TDB is frequently used in analysis of slight difference of TDB-TT and also usually used in the solar system dynamics model. Currently in post-Newtonian equations, TCB and TDB time scales are both applicable, but the second length coefficient ratio of TDB and TCB is L_B.

2.1.2.5 Time System in Pulsar-Based Navigation

In pulsar-based navigation, what is measured directly on the spacecraft is proper time dependent on the gravitational field where the spacecraft is and its motion state, while one certain coordinate time shall be used to describe the motion and phase propagation model. Hence, the measured proper time needs to be transformed to coordinate time. The delay of proper time relative to coordinate time is known as Einstein delay.

Einstein delay can be obtained based on the integral of Eq. 2.6, in which the expressions of U and v are related to the selected coordinate time. However, in specific contexts, more simplified expressions can be derived. The following is only dedicated to discussing the two-body issue, which has an accuracy high enough in the vast majority of cases.

(1) **Near-earth orbit**

For an Earth-centered orbit, proper time can be transformed to TCG at first, and then be transformed to TCB, TDB and TT as needed.

For transformation from proper time to TCG,

$$\begin{cases} U = -\frac{\mu_\oplus}{r_\oplus} \\ v_\oplus^2 = \mu_\oplus\left(\frac{2}{r_\oplus} - \frac{1}{a}\right) \end{cases} \tag{2.12}$$

where $\mu_\oplus$ is the earth gravitational constant, r is the distance from the spacecraft to the geocenter, v is the velocity of spacecraft and a is the semi-major axis of orbit.

Substituting Eq. (2.12) into Eq. 2.6 obtains

$$\frac{d\tau}{dt} = 1 - \frac{\mu_\oplus}{2c^2}\left(\frac{4}{r_\oplus} - \frac{1}{a}\right). \tag{2.13}$$

Now, we calculate the accumulated delay δt in an orbital period. Substitute $dE = \frac{a}{r}ndt$ into Eq. 2.13 and then integrate it to obtain:

$$\begin{aligned} \delta t &= -\frac{\mu_\oplus}{2c^2}\int_0^T\left(\frac{4}{r_\oplus} - \frac{1}{a}\right)dt \\ &= -\frac{\mu_\oplus}{2c^2}\int_0^{2\pi}\left(\frac{3}{an} + \frac{e}{an}\cos E\right)dE = -\frac{3\pi}{c^2}\sqrt{a\mu_\oplus} \end{aligned} \tag{2.14}$$

where n is the mean angular velocity of orbit and E is the eccentric anomaly.

Assume the mean scale transformation factor is

$$L_S = \frac{|\delta t|}{T} = \frac{3\mu_\oplus}{2a}\frac{1}{c^2}. \tag{2.15}$$

Generally, Earth-orbiting spacecraft are in orbits with small eccentricity and have short periods, so the clock frequency can be adjusted directly with the use of the above scale transformation factor; thus, the time measured on the spacecraft is TCG. Similarly, the time measured directly on the spacecraft can be TT through the frequency adjustment.

(2) **Deep space orbit**

For a deep space orbit, if only the effects of solar gravity are considered, then

$$\frac{d\tau}{dt} = 1 - \frac{\mu_\odot}{2c^2}\left(\frac{4}{r_\odot} - \frac{1}{a}\right). \tag{2.16}$$

The corresponding coordinates at this moment is TCB, where $\mu_\odot$ is the solar gravitational constant and r is the distance from the spacecraft to the SSB.

Assume that the accumulated delay between t_1 and t_2 is δt, thus

$$\begin{aligned}\delta t &= \int_{t_1}^{t_2} -\frac{\mu_\odot}{2c^2}\left(\frac{4}{r_\odot} - \frac{1}{a}\right)dt \\ &= -\frac{\mu_\odot}{2c^2}\int_{E_1}^{E_2}\left(\frac{3}{an} + \frac{e}{an}\cos E\right)dE \\ &= -\frac{\mu_\odot}{2c^2}\left[\frac{3}{an}(E_2 - E_1) + \frac{e}{an}(\sin E_2 - \sin E_1)\right]\end{aligned} \tag{2.17}$$

Substituting $(E_2 - e\sin E_2) - (E_1 - e\sin E_1) = \sqrt{\frac{\mu_\odot}{a^3}}(t_2 - t_1)$ into Eq. (2.17) yields

$$\delta t = -\frac{\mu_\odot}{2c^2}\left[\frac{4}{an}(E_2 - E_1) - \frac{1}{a}(t_2 - t_1)\right]. \tag{2.18}$$

If a higher accuracy is required or when the investigated spacecraft approaches a certain major planet, a more complicated gravitational environment needs to be considered. Under certain assumptions, a similar analytical solution can also be derived, which will not be explained here for simplicity.

2.2 Timing Model

In the pulsar timing analysis, timing model includes two components: time and phase model and time transfer model, where the time and phase model is to predict the pulsar phase at the inertial system (Fig. 2.1), and time transfer model is to transfer the measured pulse TOA to the SSB, correcting time delay caused by the geometrical and general relativistic effect.

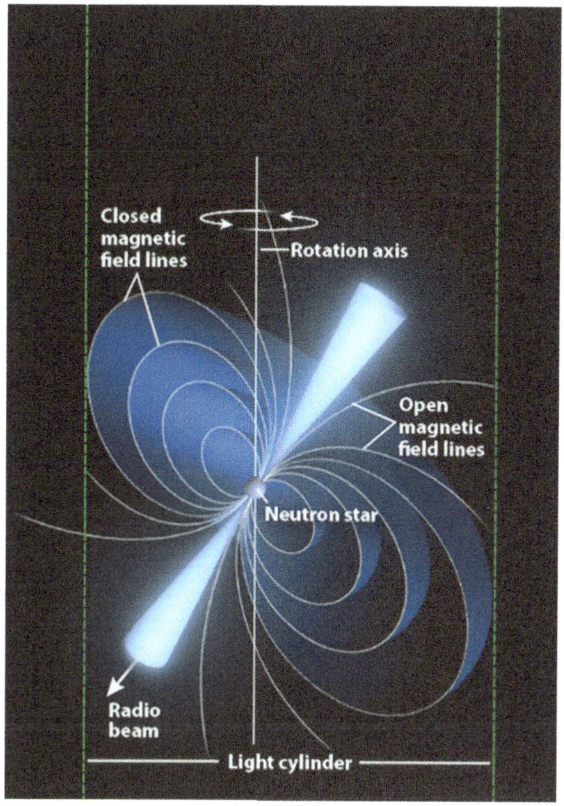

Fig. 2.1 Pulsar lighthouse model (from ESA/ATG Medialab)

2.2.1 Time and Phase Model

At SSB, the phase of pulsar can be expanded as a Taylor series around reference epoch t_0, and the time and phase model is [6]

$$\phi(t) = \phi(t_0) + \sum_{n=0}^{n=+\infty} \frac{1}{n!} f^{(n)} (t - t_0)^{n+1} \tag{2.19}$$

where $\phi(T_0)$ is the phase of pulsar signal at T_0 and $f^{(n)}$ is the nth derivative of spinning frequency of pulsar.

On the premier that the parameters in Eq. (2.19) are accurate, Eq. (2.19) can predict pulse phase at arbitrary time in the inertial system. However, limited by the current status of astronomical measurement, the derivatives of pulsar spinning frequency cannot be accurately determined, bringing out timing noise. Moreover, there are glitches in pulsars, which occur more frequently in normal or young pulsars, causing that the spinning frequency of pulsar suddenly change. During the glitch, Eq. (2.19) completely fails. Currently, there are various explanations of timing noise and of glitch, but there is not a universal theory [7, 8].

The Jodrell Bank observatory in UK keeps observing PSR B0531+21 for a long time, and publishes the ephemeris of PSR B0531+21 every month. Figures 2.2 and 2.3 show the spinning frequency and 1st time derivative of spinning frequency of PSR B0531+21 between 1988 and 2015. It can be seen from Fig. 2.3, there are numerous glitches over 17 years, and the durations of glitches are different.

Chapter 4 will discuss the impact of timing noise. As there is few public data of glitch, this book will not analyze the impact of glitch.

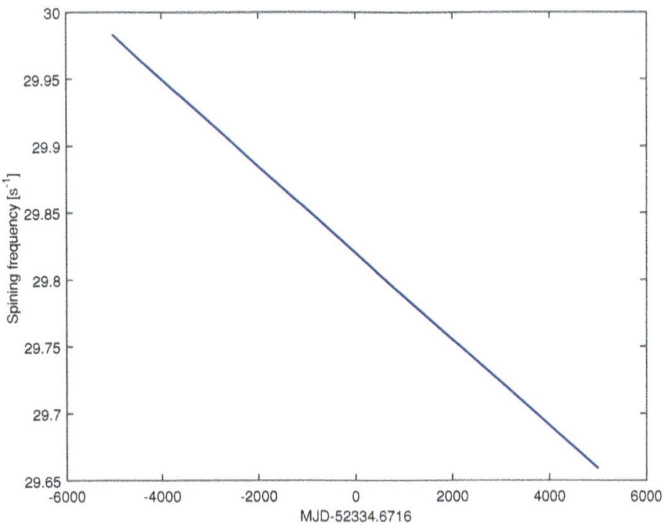

Fig. 2.2 Changes of PSR B0531+2 spinning frequency

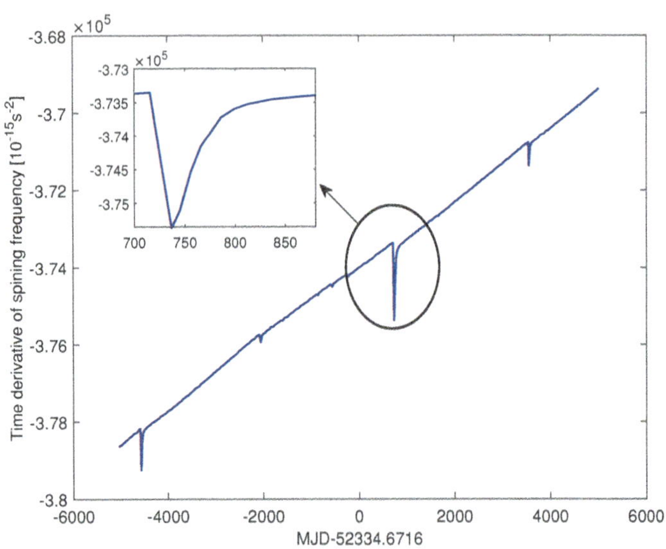

Fig. 2.3 Time derivative of spinning frequency of PSR B0531+21

2.2.2 *Time Transfer Model*

The X-ray pulsar-based navigation method works by comparing the pulse TOA at the spacecraft and that at the SSB. As the solar system is a Riemannian space, the relationship between the measured and the predicted TOA has to consider the general relativistic effect. This subsection will take rotation-powered pulsar as an example, and give the time transfer model that connects the measured TOA and the predicted one.

2.2.2.1 Accurate Time Transfer Model

In consideration of effects such as annual parallax, Roamer delay, dispersion delay and Shapiro time delay, supposing that the time when the pulse arrives at the detector is t_{sc} and the time when the pulse arrives at the vacuum SSB is d_{SSB}, the corresponding time transfer model can be written as follows [9]

$$
\begin{aligned}
d_{SSB} - t_{SC} = {} & \frac{1}{c}|\boldsymbol{D} - \boldsymbol{b}| - \frac{1}{c}|\boldsymbol{D} - \boldsymbol{p}| + \sum_{k=1}^{N} \frac{2\mu_k}{c^3} \ln\left|\boldsymbol{n}_{sc} \cdot \boldsymbol{p}_k + \|\boldsymbol{p}_k\|\right| \\
& - \frac{2\mu_S^2}{c^5 D_y^2} \left\{ \begin{aligned} & |\boldsymbol{D} - \boldsymbol{p}| \left[\left(\frac{\boldsymbol{n}_{sc} \cdot \boldsymbol{D}}{\|\boldsymbol{D}\|} \right)^2 + 1 \right] + 2(\boldsymbol{n}_{sc} \cdot \boldsymbol{D}) \left(\frac{\|\boldsymbol{p}\|}{\|\boldsymbol{D}\|} - 1 \right) \\ & + D_y \left[\arctan\left(\frac{\boldsymbol{n}_{sc} \cdot \boldsymbol{D}}{D_y} \right) - \arctan\left(\frac{\boldsymbol{n}_{sc} \cdot \boldsymbol{p}}{D_y} \right) \right] \end{aligned} \right\}
\end{aligned}
\tag{2.20}
$$

where $\boldsymbol{D}$ is the position of the pulsar relative to the Sun, $\boldsymbol{b}$ is the position of SSB relative to the Sun, $\boldsymbol{n}_{sc}$ is the directional vector of the spacecraft pointing to the pulsar, $\boldsymbol{p}$ is the position of the spacecraft relative to the Sun, N is the number of celestial bodies considered, μ_k is the gravitational constant of kth celestial body, $\boldsymbol{p}_k$ is the position of the spacecraft relative to the kth planet, μ_S is the solar gravitational constant, D_y is the component of $\boldsymbol{D}$ in the y direction. The location relationship among the spacecraft and celestial bodies is given in Fig. 2.4.

It is worth noting that the pulse TOA calculated at the spacecraft is proper time, so it needs to be transferred to the coordinate time based on Eq. (2.6) first, and then time transformation is performed based on Eq. (2.20).

2.2.2.2 Time Transfer Model for Navigation

At first, the bending effect is neglected because its influence is less than 1 ns.

The position change of the pulsar can be approximately represented like this: the pulsar moves from position $\boldsymbol{D}_0$ where it receives the 0th pulsar at t_{T_0} to position $\boldsymbol{D}_N$ where it receives the Nth pulse at t_{T_N}. Assuming that the pulsar moves at a constant speed V, we have

$$
\boldsymbol{D}_N = \boldsymbol{D}_0 + \boldsymbol{V}(t_{T_N} - t_{T_0}) \approx \boldsymbol{D}_0 + \boldsymbol{V}(t_{SC_N} - t_{SC_0}) = \boldsymbol{D}_0 + \boldsymbol{V}\Delta t_N \tag{2.21}
$$

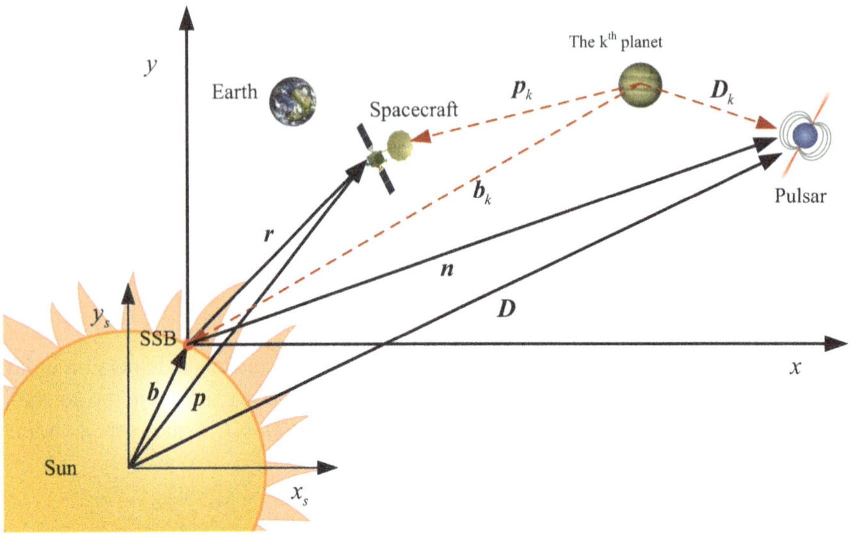

Fig. 2.4 Position relationship between spacecraft and celestial body

where t_{SC_N} and t_{SC_0} are the time when the spacecraft receives the Nth and 0th pulses respectively.

Assuming that the directional vector of the pulsar in the solar system stays the same, thus

$$\boldsymbol{n}_{sc} \approx \boldsymbol{n} \approx \boldsymbol{D}_0 / D_0. \tag{2.22}$$

Substitute Eqs. (2.21) and (2.22) into Eq. (2.20) and expand the first and second terms in Eq. (2.20) to the second order according to the Taylor series, neglect the influence of the term $O(D_0^3)$ and suppose that $\boldsymbol{r}$ indicates the position of the spacecraft relative to the SSB. Thus

$$\begin{aligned} d_{SSB} - t_{SC} \\ = \frac{1}{c} & \left\{ \begin{array}{l} \boldsymbol{n}\cdot\boldsymbol{r} - \frac{\|\boldsymbol{r}\|^2}{2D_0} + \frac{(\boldsymbol{n}\cdot\boldsymbol{r})^2}{2D_0} + \frac{\boldsymbol{r}\cdot\boldsymbol{V}\Delta t_N}{D_0} - \frac{(\boldsymbol{n}\cdot\boldsymbol{V}\Delta t_N)(\boldsymbol{n}\cdot\boldsymbol{r})}{D_0} - \frac{(\boldsymbol{b}\cdot\boldsymbol{r})}{D_0} + \frac{(\boldsymbol{n}\cdot\boldsymbol{b})(\boldsymbol{n}\cdot\boldsymbol{r})}{D_0} \\ + \frac{(\boldsymbol{n}\cdot\boldsymbol{V}\Delta t_N)}{D_0^2}\left[\frac{r^2}{2} - (\boldsymbol{r}\cdot\boldsymbol{V}\Delta t_N) + (\boldsymbol{b}\cdot\boldsymbol{r})\right] + \frac{(\boldsymbol{n}\cdot\boldsymbol{b})}{D_0^2}\left[-(\boldsymbol{b}\cdot\boldsymbol{r}) - \frac{\|\boldsymbol{r}\|^2}{2} + \boldsymbol{r}\cdot\boldsymbol{V}\Delta t_N\right] \\ + \frac{(\boldsymbol{n}\cdot\boldsymbol{r})}{D_0^2}\left[(\boldsymbol{r}\cdot\boldsymbol{V}\Delta t_N) - \frac{\|\boldsymbol{r}\|^2}{2} - (\boldsymbol{b}\cdot\boldsymbol{r}) - \frac{b^2}{2} + (\boldsymbol{b}\cdot\boldsymbol{V}\Delta t_N)\right] \end{array} \right\}. \\ & + \sum_{k=1}^{N} \frac{2\mu_k}{c^3} \ln\left|\boldsymbol{n}\cdot\boldsymbol{p}_k + \|\boldsymbol{p}_k\|\right| \end{aligned} \tag{2.23}$$

If the term of $O(D_0^2)$ is neglected, Eq. (2.23) can be simplified as

$$d_{SSB} - t_{SC} = \frac{1}{c}\left\{ \begin{array}{l} \boldsymbol{n}\cdot\boldsymbol{r} - \frac{\|\boldsymbol{r}\|^2}{2D_0} + \frac{(\boldsymbol{n}\cdot\boldsymbol{r})^2}{2D_0} + \frac{\boldsymbol{r}\cdot\boldsymbol{V}\Delta t_N}{D_0} - \frac{(\boldsymbol{n}\cdot\boldsymbol{V}\Delta t_N)(\boldsymbol{n}\cdot\boldsymbol{r})}{D_0} \\ - \frac{(\boldsymbol{b}\cdot\boldsymbol{r})}{D_0} + \frac{(\boldsymbol{n}\cdot\boldsymbol{b})(\boldsymbol{n}\cdot\boldsymbol{r})}{D_0} \end{array} \right\}$$

$$+\sum_{k=1}^{N}\frac{2\mu_k}{c^3}\ln\left|\boldsymbol{n}\cdot\boldsymbol{p}_k+\|\boldsymbol{p}_k\|\right|. \tag{2.24}$$

If the influence of proper motion of the pulsar is neglected, Eq. (2.24) can be further simplified as

$$d_{SSB}-t_{SC}=\frac{1}{c}\left[\begin{array}{l}\boldsymbol{n}\cdot\boldsymbol{r}-\frac{\|\boldsymbol{r}\|^2}{2D_0}+\frac{(\boldsymbol{n}\cdot\boldsymbol{r})^2}{2D_0}-\frac{(\boldsymbol{b}\cdot\boldsymbol{r})}{D_0}\\+\frac{(\boldsymbol{n}\cdot\boldsymbol{b})(\boldsymbol{n}\cdot\boldsymbol{r})}{D_0}\end{array}\right]+\sum_{k=1}^{N}\frac{2\mu_k}{c^3}\ln\left|\boldsymbol{n}\cdot\boldsymbol{p}_k+\|\boldsymbol{p}_k\|\right|. \tag{2.25}$$

As illustrated in Eq. (2.25), the celestial bodies' ephemeris errors, pulsar direction error, pulsar distance error, and spacecraft-borne clock error all could affect the accuracy of time transfer model. Chapter 5 will analyze the impact of these error sources.

2.3 Spacecraft Orbital Dynamics and Attitude Dynamics Models

2.3.1 Spacecraft Orbital Dynamics Model

The spacecraft orbital dynamics model can be modeled as

$$\begin{bmatrix}\dot{\boldsymbol{r}}\\ \dot{\boldsymbol{v}}\end{bmatrix}=\begin{bmatrix}\boldsymbol{v}\\ \boldsymbol{a}\end{bmatrix} \tag{2.26}$$

where $\boldsymbol{r}$ and $\boldsymbol{v}$ are the position and velocity of the spacecraft respectively. The orbital dynamics model varies as the expression of $\boldsymbol{a}$ varies.

2.3.1.1 Near-Earth Space Case

In the near-Earth space, $\boldsymbol{a}=\boldsymbol{a}_{TB}+\boldsymbol{a}_{NS}+\boldsymbol{a}_T+\boldsymbol{a}_{SP}+\boldsymbol{a}_{ST}+\boldsymbol{a}_F$ is the acceleration of the spacecraft, where $\boldsymbol{a}_{TB}$ is the Earth's two-body gravitational acceleration; $\boldsymbol{a}_{NS}$ is the Earth's non-spherical perturbation acceleration; $\boldsymbol{a}_T$ is the gravitational perturbation acceleration caused by the other celestial bodies; $\boldsymbol{a}_{SP}$ is the solar radiation pressure perturbation acceleration; $\boldsymbol{a}_{ST}$ is the perturbation acceleration caused by lunar solar solid tides; $\boldsymbol{a}_F$ is the atmospheric drag perturbation acceleration.

(1) **Earth's two-body gravitational acceleration**

Assuming the Earth is a homogeneous sphere, the two-body gravitational acceleration can be given as [10]

$$\boldsymbol{a}_{TB}=-\mu_E\frac{\boldsymbol{r}}{\|\boldsymbol{r}\|^3} \tag{2.27}$$

where μ_E is the earth gravitational constant.

(2) Earth's non-spherical perturbation acceleration

Since the mass distribution of Earth is actually not even, it would introduce great errors if the Earth is straightforwardly assumed as a homogeneous sphere. The Earth's non-spherical perturbation acceleration is given by:

$$\boldsymbol{a}_{NS} = \frac{\partial U_{NSE}}{\partial \boldsymbol{r}} \tag{2.28}$$

where U_{NSE} is the Earth gravitational potential. It is usually described by spherical harmonics and given by [11]

$$U_{NSE} = \frac{\mu_E}{\|\boldsymbol{r}\|}\left[1 + \sum_{n=2}^{\infty}\sum_{m=1}^{n}\left(\frac{R_e}{\|\boldsymbol{r}\|}\right)^n P_n^m(\sin\phi)(A_{n,m}\cos m\lambda + B_{n,m}\sin m\lambda)\right] \tag{2.29}$$

where R_e is the radius of Earth, ϕ and λ are the geocenter longitude and latitude, $A_{n,m}$ and $B_{n,m}$ are spherical harmonics.

(3) Gravitational perturbation acceleration caused by the other celestial bodies

Spacecraft may be influenced by gravity of the Sun, moon and other giant celestial bodies. If other celestial bodies are regarded as the mass point, the perturbation acceleration caused by the other celestial bodies can approximately be

$$\boldsymbol{a}_T = \sum_{i=1}^{n}\mu_i\left[\frac{\boldsymbol{r}_i}{\|\boldsymbol{r}_i\|^3} - \frac{\boldsymbol{r} - \boldsymbol{r}_i}{\|\boldsymbol{r} - \boldsymbol{r}_i\|^3}\right] \tag{2.30}$$

where μ_i is the gravitational constant of the ith celestial body and $\boldsymbol{r}_i$ is the position of the ith celestial body relative to the Earth.

(4) Solar radiation pressure perturbation acceleration

When the spacecraft is exposed to the Sun, some part of solar radiation energy will be absorbed, and the rest will be reflected. Such energy transformation enables the spacecraft to be exerted by the solar radiation pressure. Sunlight emissions on the surface of the spacecraft are very complicated, so for convenience, it is generally acknowledged that the radiation pressure is in the same direction as the incident sunlight and the radiation pressure perturbation acceleration applied on the mass of the spacecraft are [2]

$$\boldsymbol{a}_{SP} = -KpC_R\left(\frac{A}{m}\right)\boldsymbol{S} \tag{2.31}$$

where K is the shined factor; p is the radiation pressure applied on the black body in a certain distance from the Sun, in the unit of N/m^2; C_R is the surface reflection coefficient of the spacecraft; A is the spacecraft's cross-sectional area perpendicular to the direction of sunlight; m is the mass of the spacecraft; $\boldsymbol{S}$ is the unit direction vector from the Earth to the Sun.

(5) **Perturbation acceleration of lunar solar solid tides**

Due to the influences of the solar and lunar gravity, the Earth is subjected to elastic deformations manifested by solid tides, sea tides and atmospheric tides. The solid tide originates in elastic deformations on the land and may cause a 20–30 cm wavy amplitude of the Earth's crust. The sea tide and atmospheric tide are additional effects of the solid tide and influence the motion of the spacecraft together with the solid tide.

The spacecraft will incur the perturbation potential as follows due to the tidal deformation caused by the solar and lunar gravity [12].

$$\begin{cases} U_T = \frac{\mu_S}{\|\boldsymbol{r}_S\|^3}\frac{R_e^5}{\|\boldsymbol{r}\|^3}K_2P_2(\cos\psi_S) + \frac{\mu_L}{\|\boldsymbol{r}_L\|^3}\frac{R_e^5}{\|\boldsymbol{r}\|^3}K_2P_2(\cos\psi_L) \\ P_2(\cos\psi_S) = \frac{1}{2}\left[3\left(\boldsymbol{r}_S^T\cdot\boldsymbol{r}\right) - 1\right] \\ P_2(\cos\psi_L) = \frac{1}{2}\left[3\left(\boldsymbol{r}_L^T\cdot\boldsymbol{r}\right) - 1\right] \end{cases} \tag{2.32}$$

where K_2 is the second-order Love number, often taken as $K_2 = 0.3$; S and L indicate the sun and the moon respectively; $\boldsymbol{r}_S$ and $\boldsymbol{r}_L$ indicate the positions of the Sun and the moon relative to the Earth respectively; ψ_S is the angle between the positions of the Sun and the spacecraft; ψ_L is the angle between the positions of the moon and the spacecraft.

Hence, the solid tide perturbation acceleration applied to the spacecraft is

$$\begin{aligned} \boldsymbol{a}_{ST} = \frac{\partial U_T}{\partial \boldsymbol{r}} &= \frac{\mu_S}{2\|\boldsymbol{r}_S\|^3}\frac{R_e^5}{\|\boldsymbol{r}\|^4}K_2\left\{\left[3 - 15(\boldsymbol{r}_S\cdot\boldsymbol{r})^2\right]\boldsymbol{r} + 6(\boldsymbol{r}_S\cdot\boldsymbol{r})^2\boldsymbol{r}_S\right\} \\ &+ \frac{\mu_L}{2\|\boldsymbol{r}_L\|^3}\frac{a_e^5}{\|\boldsymbol{r}\|^4}K_2\left\{\left[3 - 15(\boldsymbol{r}_L\cdot\boldsymbol{r})^2\right]\boldsymbol{r} + 6(\boldsymbol{r}_L\cdot\boldsymbol{r})^2\boldsymbol{r}_L\right\}. \end{aligned} \tag{2.33}$$

(6) **Atmospheric drag perturbation acceleration**

The spacecraft on the orbit within the atmosphere will be influenced by the air drag during operation. The air drag acceleration applied to the spacecraft is given by

$$\boldsymbol{a}_F = -\frac{1}{2}C_D\left(\frac{A}{m}\right)\rho v'\boldsymbol{v}' \tag{2.34}$$

where C_D is the drag coefficient, ρ is the atmospheric density in the space where the satellite is, A/m is the ratio of the spacecraft's windward area to its mass and $\boldsymbol{v}'$ is its windward velocity.

2.3.1.2 Deep Space Case

For the deep space explorers, their position and velocity are defined in the Sun-centered inertial system and $\boldsymbol{a}$ in Eq. (2.26) is

$$\boldsymbol{a} = -\frac{\mu_S}{\|\boldsymbol{r}\|^3}\boldsymbol{r} + \sum_{i=1}^{n_p} \mu_i \left(\frac{\boldsymbol{r}_{ri}}{\|\boldsymbol{r}_{ri}\|^3} - \frac{\boldsymbol{r}_{pi}}{\|\boldsymbol{r}_{pi}\|^3} \right) + \boldsymbol{a}_{H.O.T} \tag{2.35}$$

where μ_i is the gravitational constant of the ith perturbation celestial body; $\boldsymbol{r}_{ri}$ is the position of the ith perturbation celestial body relative to the Sun; $\boldsymbol{r}_{pi}$ is the position of the deep space explorer relative to the ith perturbation celestial body; $\boldsymbol{a}_{H.O.T}$ is the high-order perturbation acceleration such as solar radiation pressure perturbation acceleration.

2.3.2 *Spacecraft Attitude Dynamics Model*

2.3.2.1 Attitude Description

There are many parameters that could represent the attitude of spacecraft, such as Euler angles [13–15], quaternion [14–16], modified Rodrigues parameters [17, 18] (MRPs) and direction cosine matrixes [19]. The Euler angle is commonly used for describing the attitude of a rigid body and is a set of minimal realizations of attitude description. However, it involves massive trigonometric operations, so it is not applicable in the case of a high real-time requirement. The quaternion method, covering the shortage of the Euler angle method, has been frequently used in calculating the attitude of a spacecraft. In the kinematical equation represented by quaternions, there are no singular points. But the quaternion method has its constraint, and it is not the minimal realization of attitude description. When quaternions are employed, some problems might occur if there is not a treatment on the constraint of quaternions. Hence modified Rodrigues parameters are proposed.

(1) **Quaternion**

The quaternion is defined as

$$\boldsymbol{q} = q_1 + q_2\boldsymbol{i} + q_3\boldsymbol{j} + q_4\boldsymbol{k} = \left[\boldsymbol{\xi}^T \; q_1\right]^T \tag{2.36}$$

where $\boldsymbol{\xi} = \left[q_2 \; q_3 \; q_4\right]^T = \boldsymbol{e}\sin(\vartheta/2)$, $q_1 = \cos(\vartheta/2)$, $\boldsymbol{e}$ is the rotation axis, ϑ is the rotation angle, q_1 is the scalar part of $\boldsymbol{q}$ and $\boldsymbol{\xi}$ is the vector part of $\boldsymbol{q}$.

The addition operation of quaternions is defined to be the sum of two elements corresponding to two quaternions. The multiplication of two quaternions $\boldsymbol{p}$ and $\boldsymbol{q}$ is defined as follows:

$$\boldsymbol{p} \otimes \boldsymbol{q} = \begin{bmatrix} p_1 & -p_4 & p_3 & p_2 \\ p_4 & p_1 & -p_2 & p_3 \\ -p_3 & p_2 & p_1 & p_4 \\ -p_2 & -p_3 & -p_4 & p_1 \end{bmatrix} \begin{bmatrix} q_2 \\ q_3 \\ q_4 \\ q_1 \end{bmatrix}. \tag{2.37}$$

Suppose that the vector $\boldsymbol{r}$ rotates around an axis $\boldsymbol{e}$ by an angle of ϑ to arrive at $\boldsymbol{r}'$, as shown in Fig. 2.5. Take $\boldsymbol{e}$ as a unit vector. Thus

$$\boldsymbol{r}' = \boldsymbol{r}\cos\vartheta + (1 - \cos\vartheta)(\boldsymbol{r}\cdot\boldsymbol{e})\boldsymbol{e} + (\boldsymbol{e}\times\boldsymbol{r})\sin\vartheta. \tag{2.38}$$

Based on the definition of the quaternion, it can be easily derived that

$$\boldsymbol{r}' = \boldsymbol{q}\boldsymbol{r}\boldsymbol{q}^{-1} \tag{2.39}$$

where $\boldsymbol{q}^{-1} = \left[-\boldsymbol{\xi}^{\mathrm{T}}\ q_1\right]^T$. $\boldsymbol{r}$ and $\boldsymbol{r}'$ are expressed by quaternions: $\boldsymbol{r} = [x\ y\ z\ 0]^T$ and $\boldsymbol{r}' = \left[x'\ y'\ z'\ 0\right]^T$. Expand Eq. (2.39) to obtain

$$\begin{bmatrix} x' \\ y' \\ z' \end{bmatrix} = \begin{bmatrix} q_1^2 + q_2^2 - q_3^2 - q_4^2 & -2q_1q_4 + 2q_2q_3 & 2q_1q_3 + 2q_2q_4 \\ 2q_1q_4 + 2q_2q_3 & q_1^2 - q_2^2 + q_3^2 - q_4^2 & -2q_1q_2 + 2q_3q_4 \\ -2q_1q_3 + 2q_2q_4 & 2q_1q_2 + 2q_3q_4 & q_1^2 - q_2^2 - q_3^2 + q_4^2 \end{bmatrix} \begin{bmatrix} x \\ y \\ z \end{bmatrix}. \tag{2.40}$$

(2) **Modified Rodrigues parameters**

Rodrigues parameters are defined as

$$\boldsymbol{\Theta} = \left[\Theta_x\ \Theta_y\ \Theta_z\right]^T = \boldsymbol{e}\tan(\vartheta/2). \tag{2.41}$$

Rodrigues parameters defined above are also called Gibbs parameters. The addition operation for Rodrigues parameters is the same as that for ordinary vectors,

Fig. 2.5 Vector rotation

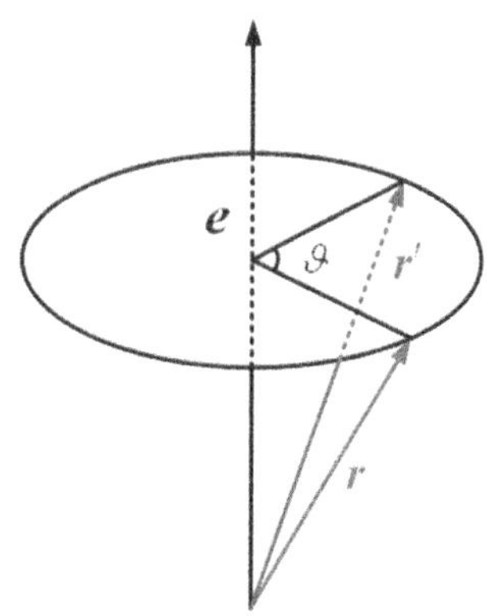

$$\boldsymbol{\Theta}_1 + \boldsymbol{\Theta}_2 = \left[\Theta_{1x} + \Theta_{2x},\ \Theta_{1y} + \Theta_{2y},\ \Theta_{1z} + \Theta_{2z}\right]^T. \tag{2.42}$$

The multiplication operation of Rodrigues parameters is defined as follows:

$$\boldsymbol{\Theta}_1 * \boldsymbol{\Theta}_2 = \frac{\boldsymbol{\Theta}_1 + \boldsymbol{\Theta}_2 + \boldsymbol{\Theta}_1 \times \boldsymbol{\Theta}_2}{1 - (\boldsymbol{\Theta}_1 \cdot \boldsymbol{\Theta}_2)}. \tag{2.43}$$

If the vector $\boldsymbol{r}$ rotates around an axis $\boldsymbol{e}$ by ϑ to arrive at $\boldsymbol{r}'$, $\boldsymbol{r}'$ can be given by

$$\boldsymbol{r}' = \frac{2}{1+\Theta^2}[\boldsymbol{\Theta} \times \boldsymbol{r} + \boldsymbol{\Theta}(\boldsymbol{\Theta} \cdot \boldsymbol{r})] + \frac{1-\Theta^2}{1+\Theta^2}\boldsymbol{r} \tag{2.44}$$

where Θ is the module of $\boldsymbol{\Theta}$ Based on the triple vector product formula, the above equation can be:

$$\boldsymbol{r}' = \boldsymbol{r} + \frac{2}{1+\Theta^2}[\boldsymbol{\Theta} \times \boldsymbol{r} + \boldsymbol{\Theta} \times (\boldsymbol{\Theta} \times \boldsymbol{r})]. \tag{2.45}$$

Based on the definition of Rodrigues parameters, when the equivalent rotation angle tends to be $\pm 180°$, $\Theta \to \infty$, so they are not applicable to describe the general attitude kinematics; however, when the equivalent rotation angle changes within a range far lower than $\pm 180°$, Rodrigues parameters are applicable to attitude description. To cover the shortage of Rodrigues parameters, scholars, by projecting quaternions on the three-dimensional hyper-plane, proposed the method of describing attitude kinematics with MRPs,

$$\boldsymbol{\Theta} = \frac{\boldsymbol{\xi}}{1+q_1} = \frac{\boldsymbol{e}\sin(\vartheta/2)}{1+\cos(\vartheta/2)} = \boldsymbol{e}\tan\frac{\vartheta}{4}. \tag{2.46}$$

Obviously, if attitude kinematics is described with MRPs, when the equivalent rotation angle tends to be $\pm 360°$, $\Theta \to \infty$, so the largest equivalent rotation angle is $(-360°, 360°)$.

2.3.2.2 Equation of Quaternion Attitude Kinematics and Equation of Euler Attitude Dynamics

The attitude kinematics equation describes the relationship between the change rate of attitude parameters and the angular velocity of the spacecraft while the attitude dynamics equation describes the relationship between the angular momentum of the spacecraft and the external applied moment. Based on quaternions, this subsection gives the quaternion kinematics equation and the Euler dynamics equation.

(1) **Equation of quaternion attitude kinematics**

Suppose that $\boldsymbol{q}$ is the rotation quaternion from the spacecraft body coordinate system to the inertial coordinate system. The attitude kinematics equation expressed by quaternions is

$$\dot{\boldsymbol{q}} = \frac{1}{2}\boldsymbol{\Omega}\boldsymbol{q} \tag{2.47}$$

where $\boldsymbol{\Omega}$ is the skew-symmetric matrix of the rotating angular velocity from the body coordinate system to the J2000 coordinate system, given by

$$\boldsymbol{\Omega} = \begin{bmatrix} 0 & \omega_z^b & -\omega_y^b & \omega_x^b \\ -\omega_z^b & 0 & \omega_x^b & \omega_y^b \\ \omega_y^b & -\omega_x^b & 0 & \omega_z^b \\ -\omega_x^b & -\omega_y^b & -\omega_z^b & 0 \end{bmatrix}. \tag{2.48}$$

(2) **Equation of Euler attitude dynamics**

In the spacecraft body coordinate system, the equation of Euler attitude dynamics can be established as

$$\boldsymbol{I}\dot{\boldsymbol{\omega}} + \boldsymbol{\omega} \times (\boldsymbol{I}\boldsymbol{\omega}) = \boldsymbol{M} \tag{2.49}$$

where $\boldsymbol{I}$ is the inertia matrix of the spacecraft, $\boldsymbol{\omega}$ is its angular velocity and $\boldsymbol{M}$ is the externally applied moment.

2.4 X-Ray Pulsar-Based Spacecraft Positioning

2.4.1 Basic Principle

The X-ray pulsar-based spacecraft positioning works by comparing the pulse TOA at the spacecraft and the TOA of the same pulse at the SSB predicted by the time and phase model. The difference between the measured and predicted pulse TOAs reflects the distance of the spacecraft relative to the SSB (Fig. 2.6). And then, the position of spacecraft can be geometrically determined by nonlinear least square algorithm when three or more pulsars are observed at the same time. When the orbital dynamics of spacecraft is taken into consideration, the position can also be determined by means of nonlinear filter, such as extended Kalman filter and unscented Kalman filter, and by sequentially observing pulsars.

Equation (2.25) is the very measurement model of X-ray pulsar-based spacecraft positioning when observing one pulsar. Assuming there are N pulsars observed at the same time, the measurement model can be presented in a matrix form of

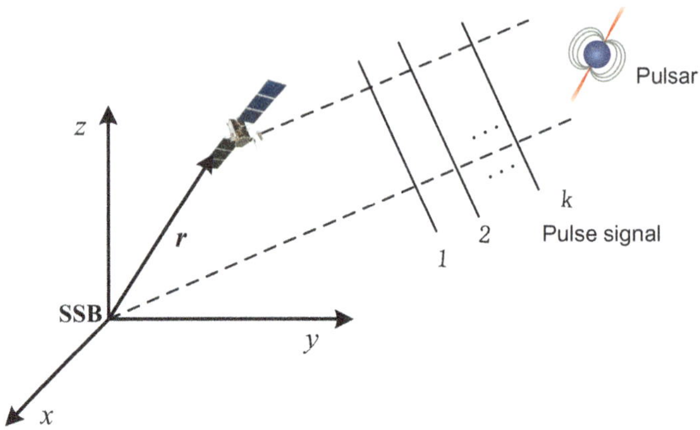

Fig. 2.6 Principle of X-ray pulsar-based navigation

$$\boldsymbol{Z} = \boldsymbol{h}(\boldsymbol{r}) + \boldsymbol{v} \tag{2.50}$$

where

$$\mathbf{Z} = \left[t_{SSB}^1 - t_{SC}^1 \cdots t_{SSB}^i - t_{SC}^i \cdots t_{SSB}^N - t_{SC}^N \right]^T, \tag{2.51}$$

$$\boldsymbol{h}(\boldsymbol{r}) = \left[\boldsymbol{h}^1(\boldsymbol{r}) \cdots \boldsymbol{h}^i(\boldsymbol{r}) \cdots \boldsymbol{h}^N(\boldsymbol{r}) \right]^T, \tag{2.52}$$

$$\begin{aligned}\mathbf{h}^i(\mathbf{r}) = \frac{1}{c}\left[\mathbf{n}^{iT}\mathbf{r} - \frac{\|\mathbf{r}\|^2}{2D_0^i} + \frac{\left(\mathbf{n}^{iT}\mathbf{r}\right)^2}{2D_0^i} - \frac{\left(\mathbf{b}^T\mathbf{r}\right)}{D_0^i} + \frac{\left(\mathbf{n}^{iT}\mathbf{b}\right)\left(\mathbf{n}^{iT}\mathbf{r}\right)}{D_0^i} \right] \\ + \sum_{k=1}^{N} \frac{2\mu_k}{c^3} \ln\left|\mathbf{n}^{iT}\mathbf{p}_k + \|\mathbf{p}_k\|\right|.\end{aligned} \tag{2.53}$$

In Eq. (2.50), $\boldsymbol{v}$ is the measurement noise which can be modeled as a zero-mean Gauss white noise with a standard deviation of

$$\sigma_{TOA} = \frac{W\sqrt{[B_X + F_X(1 - p_f)]d + F_X p_f}}{2F_X p_f \sqrt{AT_m}} \tag{2.54}$$

where W is the width of the pulse, B_X and F_X are the X-ray radiation flux of the background and pulsar respectively, p_f is the pulsed fraction of the pulsar, d equals W/P with P of the pulse period, A is the effective area of X-ray sensor, and T_m is the duration of the pulsar observation.

From intuitive analysis of Eq. (2.50), the measurement model of the X-ray pulsar-based spacecraft positioning involves the integer ambiguity of pulse signals. Unless the spacecraft is "completely lost" in the solar system and does not know any information about itself, the spacecraft can obtain relatively accurate initial state for navigation through the ground station or other navigation means. The minimum rotation period of a millisecond pulsar is 1 ms and the corresponding light travel distance is about 300 km. Also, the current navigation means mostly provide an initial navigation solution which is superior to 300 km. In addition, within one pulsar observation period, the spacecraft's orbital dynamics propagation accuracy is much higher than 300 km. Hence, the integer ambiguity of pulse signals can be neglected in the measurement model of X-ray pulsar spacecraft navigation.

2.4.2 Working Flow

The X-ray pulsar-based spacecraft positioning system mainly consists of an X-ray detector, spacecraft-borne atomic clock, spacecraft-borne computational device, navigation model algorithms database and pulsar model database. The workflow of the whole system is given in Fig. 2.7.

First of all, when photons arrive at the X-ray sensors, the arrival times are recorded by the spacecraft-borne atomic clock, and are transmitted into the spacecraft-borne computational device. And then, a pulse TOA is extracted out of the recorded photon TOAs. Moreover, the positioning can be fulfilled by employing the measured pulse TOA as well as time and phase model.

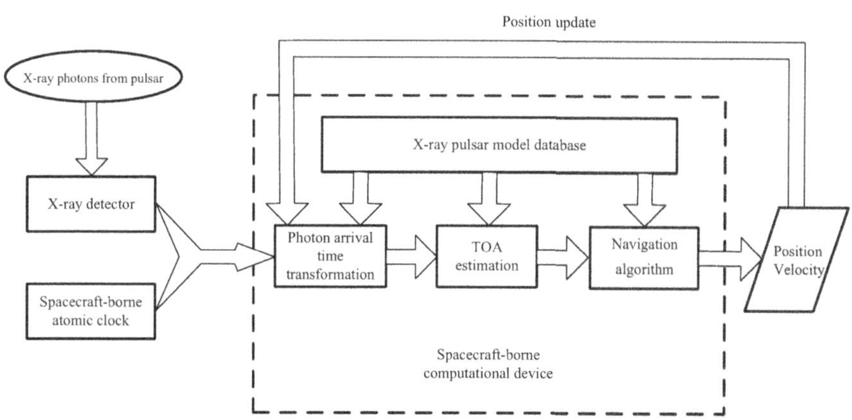

Fig. 2.7 Basic scheme of X-ray pulsar-based spacecraft positioning system

2.4.3 *Analysis on the X-Ray Detector Configuration Scheme*

Currently, the X-ray pulsar-based spacecraft positioning system works mainly under the assumption that the investigated spacecraft simultaneously carries 3 sensors with large area. However, limited by the loading capability and power consumption of the spacecraft, the assumption is not easy to realize. Thus, X-ray pulsar-based spacecraft positioning using one sensor is a good idea. This subsection will compare the positioning performance of two schemes, including the one sensor scheme and the small-area sensor scheme.

2.4.3.1 One Sensor Scheme

As the observability analysis shown, the positioning performance of X-ray pulsar-based spacecraft positioning system would not be satisfactory if the spacecraft keeps observing one pulsar over the whole navigation period. We assume the X-ray sensor locates on a platform whose direction is adjustable. The X-ray sensor could point

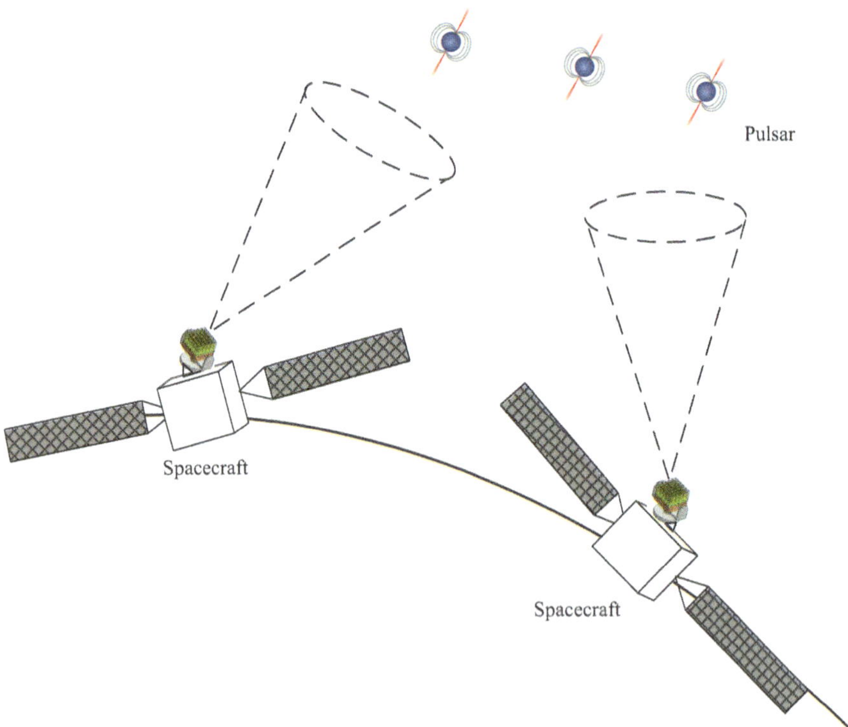

Fig. 2.8 Singular detector configuration scheme

to different pulsars according to given commands. Figure 2.8 shows a conceptual description of the scheme.

2.4.3.2 Small-Area Sensor Scheme

For the purpose of reducing the loading and power consumption of the X-ray pulsar-based spacecraft positioning system, small-area sensor scheme is also a feasible scheme besides the one sensor scheme.

Compared with one sensor scheme, the navigation system carrying multiple sensors can receive pulsar signals from multiple directions and obtain more measurement information at one navigation epoch. Theoretically, an increase in measurement information may enhance the positioning performance. Nevertheless, as shown in Eq. (2.54), the measurement accuracy of X-ray pulsar-based spacecraft positioning system is dependent not only on the detector resolution but the detector area and measurement duration.

From Eq. (2.54), when the effective area of the detector is reduced to A/n, the pulse TOA measurement accuracy is reduced to $\sigma_{TOA}\sqrt{n}$ accordingly. To simplify the comparison, the following part only analyzes the case where the spacecraft carries 3 small-area sensors.

2.4.3.3 Comparison of Two Schemes

The following part, through simulation analysis, compares the positioning performances of one sensor scheme and small-area sensor scheme.

The navigation duration lasts for 7 days, and the observation duration of pulsar is 30 min. Assume that the investigated satellite is on the geosynchronous orbit, with an initial position error of 1 km and initial velocity error of 1 m/s. The angular position error of pulsars is 0.1 mas. The area of the sensor used for one sensor scheme is 1 m^2, and the small-area sensor scheme employs 3 sensors with area of 0.33 m^2.

Table 2.1 gives the flux of navigation pulsars and the pulse TOA accuracy obtained through 30 min observation by the 1 m^2 sensor. Figure 2.9 gives the comparison between the two schemes.

From Fig. 2.9, both the convergence rate and navigation accuracy of the one sensor scheme are inferior to those of the small-area sensor scheme. The reason is that the

Table 2.1 Information about pulsar selected for navigation

Navigation pulsar	Flux/($ph/m^2/s$)	Pulse TOA measurement accuracy/μs
PSR B1509−58	162	2.86
PSR B1821−24	1.93	2.42
PSR B1937+21	0.499	3.52

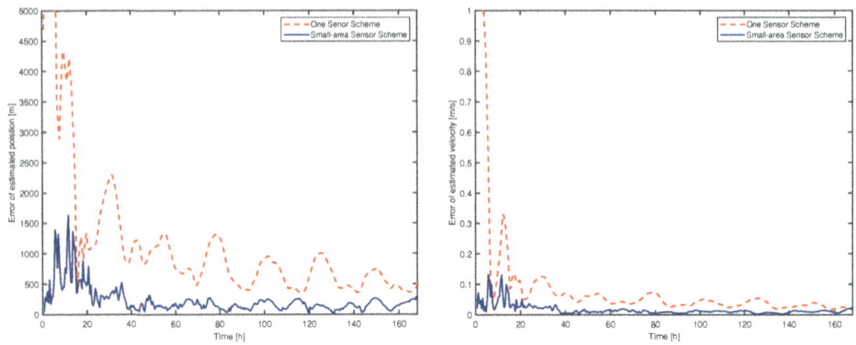

Fig. 2.9 Comparison of navigation schemes

small-area sensor scheme can provide navigation service by simultaneously using measurement information in multiple directions. However, small-area sensor scheme requires spacecraft to load supporting devices more than one sensor scheme. Thus, it should be analyzed in a comprehensive way to find out which scheme is more appropriate to practical applications.

2.5 X-Ray Pulsar-Based Spacecraft Time Keeping

2.5.1 Basic Principle

If the position of spacecraft has been well determined by the other navigation methods, the onboard clock error can be estimated by observing pulsars. In this case, we have

$$d_{SSB} = t_{sc} - \delta t + \boldsymbol{h}^i(\boldsymbol{r}) \tag{2.55}$$

where δt is the clock error and $\boldsymbol{h}^i(\boldsymbol{r})$ is defined in Eq. (2.53).

Thus, the measurement for time-keeping is

$$\Delta t = d_{SSB} - t_{sc} - \boldsymbol{h}^i(\boldsymbol{r}) = \delta t. \tag{2.56}$$

2.5.2 System Equation

The performance of the spacecraft-borne clock can be simulated by a three-state polynomial process. The model and variance of the discrete process are:

$$\begin{bmatrix} x_1(t_{k+1}) \\ x_2(t_{k+1}) \\ x_3(t_{k+1}) \end{bmatrix} = \boldsymbol{\Phi}(\tau)\begin{bmatrix} x_1(t_k) \\ x_2(t_k) \\ x_3(t_k) \end{bmatrix} + \begin{bmatrix} \omega_1(k) \\ \omega_2(k) \\ \omega_3(k) \end{bmatrix} \tag{2.57}$$

$$\begin{aligned} \mathbf{Q}_k(\tau) &= E\left[\boldsymbol{\omega}(k)\boldsymbol{\omega}(k)^T\right] \\ &= \begin{bmatrix} q_1\tau + \frac{1}{3}q_2\tau^3 + \frac{1}{20}q_3\tau^5 & \frac{1}{2}q_2\tau^2 + \frac{1}{8}q_3\tau^4 & \frac{1}{6}q_3\tau^3 \\ \frac{1}{2}q_2\tau^2 + \frac{1}{8}q_3\tau^4 & q_2\tau + \frac{1}{3}q_3\tau^3 & \frac{1}{2}q_3\tau^2 \\ \frac{1}{6}q_3\tau^3 & \frac{1}{2}q_3\tau^2 & q_3\tau \end{bmatrix} \end{aligned} \tag{2.58}$$

where x_1, x_2 and x_3 are the clock error, clock error drift rate and clock error drift change rate respectively; τ is the time interval; ω_1, ω_3 and ω_2 are clock process noises with random walk properties, $\boldsymbol{\omega} = \left[\,\omega_1\ \omega_2\ \omega_3\,\right]^T$, and q_1, q_2 and q_3 are power spectral densities of continuous process noises.

And then, we have the transition matrix of clock error is

$$\boldsymbol{\Phi}(\tau) = \begin{bmatrix} 1 & \tau & \tau^2/2 \\ 0 & 1 & \tau \\ 0 & 0 & 1 \end{bmatrix}. \tag{2.59}$$

Taking the state vector as $\boldsymbol{X} = \left[\,x_1\ x_2\ x_3\,\right]^T$, the system equation of X-ray pulse-based time-keeping system consists of Eqs. (2.56) and (2.57).

$$\begin{cases} \boldsymbol{X}(k+1) = \boldsymbol{\Phi}(\tau)\boldsymbol{X}(k) + \boldsymbol{\omega}(k) \\ \boldsymbol{Y}(k) = \boldsymbol{H}\cdot\boldsymbol{X}(k) + \boldsymbol{V}(k) \end{cases} \tag{2.60}$$

$$\boldsymbol{Y}(k) = \left[\,\Delta t_1\ \cdots\ \Delta t_i\,\right]^T \tag{2.61}$$

$$\boldsymbol{H} = \begin{bmatrix} 1\ 0\ 0 \\ \vdots \\ 1\ 0\ 0 \end{bmatrix}_{K\times 3} \tag{2.62}$$

where $\boldsymbol{V}(k)$ is the system measurement noise whose variance is $\boldsymbol{R}_k(\tau)$; K is the number of X-ray pulsars observed during the time keeping process.

2.5.3 Feasibility Analysis of Time-Keeping via the Observation of One Pulsar

2.5.3.1 Observability Analysis

Time-keeping via the observation of one pulsar has a great value of engineering application. The observation of one pulsar could reduce the whole complexity of the time-keeping system, thus reducing the system cost. This subsection will analyze the feasibility of time-keeping via the observation of one pulsar by employing the observability analysis as the complete observability of a system means that each state component of the system can be completely reflected by the output.

Lemma 1 *for the linear system*

$$\sum \begin{cases} \boldsymbol{x}(k+1) = \boldsymbol{G}\boldsymbol{x}(k), k = 0, 1, 2, \ldots \\ \boldsymbol{y}(k) = \boldsymbol{C}\boldsymbol{x}(k) \end{cases}. \tag{2.63}$$

Define the judgment matrix as

$$\boldsymbol{Q}_{ok} = \begin{bmatrix} \boldsymbol{C} \\ \boldsymbol{CG} \\ \vdots \\ \boldsymbol{CG}^{n-1} \end{bmatrix}, or, \boldsymbol{Q}_{ok}^{T} = \begin{bmatrix} \boldsymbol{C}^{T} & \boldsymbol{G}^{T}\boldsymbol{C}^{T} & \cdots & (\boldsymbol{G}^{T})^{n-1}\boldsymbol{C}^{T} \end{bmatrix} \tag{2.64}$$

Thus, the sufficient and necessary condition of which the system can be completely observed is:

$$rank(\boldsymbol{Q}_{ok}) = n \tag{2.65}$$

where $\boldsymbol{x}$ is the state of n dimension, $\boldsymbol{y}$ is the observation vector of q dimension, and $\boldsymbol{G}$ and $\boldsymbol{C}$ are constant matrices of $n \times n$ and $q \times n$.

For the time-keeping system via the observation of one pulsar, the system observation equation is

$$y = \Delta t_1 = \boldsymbol{H}_1 \cdot \boldsymbol{X} \tag{2.66}$$

with $\boldsymbol{H}_1 = \begin{bmatrix} 1 & 0 & 0 \end{bmatrix}$.

The observability matrix of the system is

$$\boldsymbol{Q} = \begin{bmatrix} \boldsymbol{H}_1 \\ \boldsymbol{H}_1\boldsymbol{\Phi} \\ \boldsymbol{H}_1\boldsymbol{\Phi}^2 \end{bmatrix} = \begin{bmatrix} 1 & 0 & 0 \\ 1 & \tau & \frac{1}{2}\tau^2 \\ 1 & 2\tau & 2\tau^2 \end{bmatrix}. \tag{2.67}$$

After elementary transformation

$$rank(\boldsymbol{Q}) = rank\begin{bmatrix} 1 & 0 & 0 \\ 0 & \tau & 0 \\ 0 & 0 & \tau^2 \end{bmatrix}. \quad (2.68)$$

The time interval $\tau > 0$ and the following equation can be derived: $rank(\boldsymbol{Q}) = 3$. Thus, the time-keeping via the observation of pulsar is observable completely and is feasible theoretically.

2.5.3.2 Simulation Analysis

Assuming the initial orbital elements of the investigated satellite are as follows: the semi-major axis is 26,378.137 km, the eccentricity is 0.1, the inclination is 60°, the right ascension of the ascending node is 0°, the true anomaly at zero time is 0°, and the argument of perigee is 0°. The initial spacecraft-borne clock error is 3.5858×10^{-6} s, the clock error drift rate is 3.637979×10^{-11} s/s and the clock error drift change rate is $6.66 \times 10^{-18} s/s^2$. Based on the rubidium atomic clock model, the noise spectrum densities of the given spacecraft-borne clock are $q_1 = 1.11 \times 10^{-22}$ s^2/s, $q_2 = 2.22 \times 10^{-32}$ s^2/s^3 and $q_3 = 6.66 \times 10^{-45}$ s^2/s^5 respectively. The time-keeping results obtained by using different numbers of pulsars are given in Fig. 2.10 as well as Table 2.2.

As the above shown, the time-keeping via the observation of one pulsar converges at a speed of little lower than that of the time-keeping via the observation of three

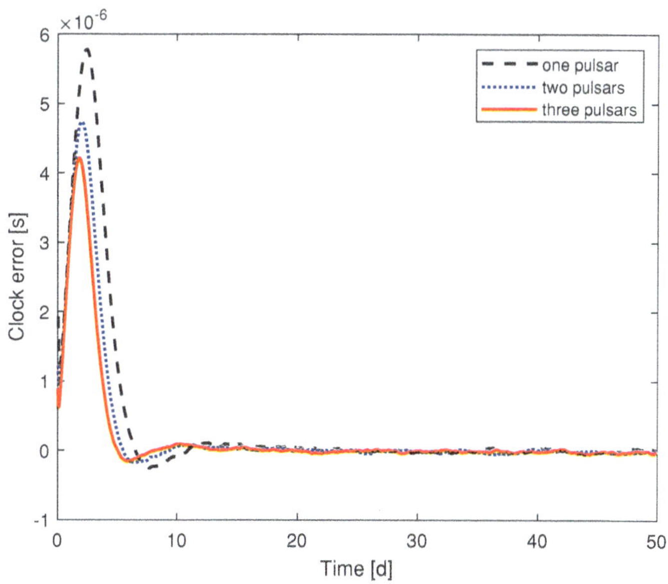

Fig. 2.10 Time-keeping result of Kalman filter

Table 2.2 Time-keeping accuracies for different numbers of pulsars [μs]

Number of pulsars	1	2	3
Estimation error of clock error	0.0589	0.0505	0.0404

pulsars, but the time-keeping accuracy is little affected by the number of pulsars. Therefore, it can be concluded that the time-keeping via the observation of one pulsar is feasible.

2.6 X-Ray Pulsar-Based Spacecraft Attitude Determination

This section first analyzes the attitude-determination principle based on direction observation and then gives introductions to some methods of realizing direction observation via the observation of pulsars.

2.6.1 Basic Principle

2.6.1.1 Two-Vector-Based Attitude Determination Principle

After processing the pulsar measurement, the direction vectors of two pulsars in the body system, $\mathbf{Z}_1^{(S)}$ and $\mathbf{Z}_2^{(S)}$ can be obtained. Assuming the right ascensions and declinations of the two pulsars are (α_1, δ_1) and (α_2, δ_2), the direction vectors of the two pulsars in the inertial system, $\mathbf{Z}_1^{(I)}$ and $\mathbf{Z}_2^{(I)}$ can be calculated out.

Thus, the relationship between the direction vectors in the inertial system and those in the body coordinate system is

$$\boldsymbol{Z}_1^{(B)} = \boldsymbol{B}_S \cdot \boldsymbol{Z}_1^{(S)}, \boldsymbol{Z}_2^{(B)} = \boldsymbol{B}_S \cdot \boldsymbol{Z}_2^{(S)}. \tag{2.69}$$

Assuming $\mathbf{Z}_1$ and $\mathbf{Z}_2$ are in different directions and $\boldsymbol{W} = \mathbf{Z}_1 \times \mathbf{Z}_2$, the components in the inertial system and the body system are

$$\boldsymbol{W}^{(I)} = \boldsymbol{Z}_1^{(I)} \times \boldsymbol{Z}_2^{(I)}, \quad \boldsymbol{W}^{(B)} = \boldsymbol{Z}_1^{(B)} \times \boldsymbol{Z}_2^{(B)}. \tag{2.70}$$

An orthogonal body system $O_{x_P y_P z_P}$ is established here. The unit vectors of all the coordinate axes are $\boldsymbol{P}$, $\boldsymbol{Q}$, $\boldsymbol{R}$, and

$$\boldsymbol{P}^{(I)} = \boldsymbol{Z}_1^{(I)}, \quad \boldsymbol{Q}^{(I)} = \boldsymbol{W}^{(I)}/\left|\boldsymbol{W}^{(I)}\right|, \quad \boldsymbol{R}^{(I)} = \boldsymbol{P}^{(I)} \times \boldsymbol{Q}^{(I)} \tag{2.71}$$

$$\boldsymbol{P}^{(B)} = \boldsymbol{Z}_1^{(B)}, \quad \boldsymbol{Q}^{(B)} = \boldsymbol{W}^{(B)}/\left|\boldsymbol{W}^{(B)}\right|, \quad \boldsymbol{R}^{(B)} = \boldsymbol{P}^{(B)} \times \boldsymbol{Q}^{(B)}. \tag{2.72}$$

Thus, the direction cosine matrices of $O_{x_p y_p z_p}$ relative to the inertial system and that relative to the body system are

$$\mathbf{M}^{(I)} = \left[\left(\mathbf{P}^{(I)}\right)^T \left(\mathbf{Q}^{(I)}\right)^T \left(\mathbf{R}^{(I)}\right)^T \right]^T,$$
$$\mathbf{M}^{(B)} = \left[\left(\mathbf{P}^{(B)}\right)^T \left(\mathbf{Q}^{(B)}\right)^T \left(\mathbf{R}^{(B)}\right)^T \right]^T. \tag{2.73}$$

Assume that the coordinates of $\boldsymbol{x}$ in the inertial system and the body system are $\boldsymbol{x}^{(I)}$ and $\boldsymbol{x}^{(B)}$ respectively

$$\boldsymbol{M}^{(I)}\boldsymbol{x}^{(I)} = \boldsymbol{M}^{(B)}\boldsymbol{x}^{(B)}. \tag{2.74}$$

And then, the transformation matrix between the inertial system to the body system is

$$\boldsymbol{M} = \left(\boldsymbol{M}^{(B)}\right)^T \boldsymbol{M}^{(I)}. \tag{2.75}$$

Then the attitude is

$$\begin{bmatrix} \varphi \\ \theta \\ \psi \end{bmatrix} = \left[\sin^{-1}(M_{23}),\ -\tan^{-1}\left(\frac{M_{13}}{M_{33}}\right),\ -\tan^{-1}\left(\frac{M_{21}}{M_{22}}\right) \right]^T. \tag{2.76}$$

Assuming that $X = \left[\varphi\ \theta\ \psi \right]^T$ is the state vector, the indirect observation equation can be established:

$$Z = \left[M_{23}\ \frac{M_{13}}{M_{33}}\ \frac{M_{21}}{M_{22}} \right]^T = \begin{bmatrix} \sin\varphi \\ -\tan\theta \\ -\tan\psi \end{bmatrix} = h(X_{K+1}, K+1). \tag{2.77}$$

Based on Eq. (2.77) and the spacecraft attitude kinematics model, the spacecraft attitude can be obtained through filtering.

2.6.1.2 Singular-Vector-Based Attitude Determination Principle

As the distribution of X-ray pulsars is scattered, it is little likely that two pulsars can be simultaneously observed. So it is necessary to study the singular-pulsar-based spacecraft attitude determination method.

Similar to the two-vector-based method, as long as the direction vector of the pulsar in the inertial system and the body system are known, the attitude matrix from the inertial system to the body system can be obtained, i.e., $\boldsymbol{Z}_B = \boldsymbol{B}_I \cdot \boldsymbol{Z}_I$. In addition, the direction vector in the inertial system, $\boldsymbol{Z}_I$, can be calculated as shown in Sect. 2.6.1.1. And then, the attitude can be estimated by employing the direction vector of the pulsar in the body system,

$$\hat{\boldsymbol{Z}}_B = \hat{\boldsymbol{B}}_I \cdot \boldsymbol{Z}_I \tag{2.78}$$

where $\hat{\boldsymbol{B}}_I$ is the attitude matrix from the inertial system to the body system, obtained through the estimated attitude.

The difference between $\boldsymbol{Z}_B$ and $\hat{\boldsymbol{Z}}_B$ reflects the offset between $\hat{\boldsymbol{B}}_I$ and $\boldsymbol{B}_I$, which could be used to estimate the attitude. Defining $\boldsymbol{e} = \boldsymbol{Z}_B - \hat{\boldsymbol{Z}}_B$, the measurement model is

$$\boldsymbol{e}_k = \boldsymbol{H}_{k/k-1} \cdot \boldsymbol{X}_k + \boldsymbol{V}_k \tag{2.79}$$

where the error attitude X_k is assumed to be the state vector and $H_{k/k-1}$ is assumed to be the observation equation,

$$X = \left[\varphi - \hat{\varphi}\ \theta - \hat{\theta}\ \psi - \hat{\psi}\right]^T = \left[\Delta\varphi\ \Delta\theta\ \Delta\psi\right]^T \tag{2.80}$$

$$\boldsymbol{H}_{k/k-1} = \left[\left.\frac{\partial \boldsymbol{B}_I}{\partial \varphi}\right|_{\hat{\varphi}_{k/k-1}} \cdot \boldsymbol{Z}_I\ \left.\frac{\partial \boldsymbol{B}_I}{\partial \theta}\right|_{\hat{\theta}_{k/k-1}} \cdot \boldsymbol{Z}_I\ \left.\frac{\partial \boldsymbol{B}_I}{\partial \psi}\right|_{\hat{\psi}_{k/k-1}} \cdot \boldsymbol{Z}_I\right] \tag{2.81}$$

where

$$\frac{\partial \boldsymbol{B}_I}{\partial \varphi} = \begin{bmatrix} -\cos\varphi\sin\theta\sin\psi & \cos\varphi\sin\theta\cos\psi & \sin\varphi\sin\theta \\ \sin\varphi\sin\psi & -\sin\varphi\cos\psi & \cos\varphi \\ \cos\varphi\cos\theta\sin\psi & -\cos\varphi\cos\theta\cos\psi & -\sin\varphi\cos\theta \end{bmatrix}, \tag{2.82}$$

$$\frac{\partial \boldsymbol{B}_I}{\partial \theta} = \begin{bmatrix} -\sin\theta\cos\psi - \sin\varphi\cos\theta\sin\psi & -\sin\theta\sin\psi + \sin\varphi\cos\theta\cos\psi & -\cos\varphi\cos\theta \\ 0 & 0 & 0 \\ \cos\theta\cos\psi - \sin\varphi\sin\theta\sin\psi & \cos\theta\sin\psi + \sin\varphi\sin\theta\cos\psi & -\cos\varphi\sin\theta \end{bmatrix}, \tag{2.83}$$

$$\frac{\partial \boldsymbol{B}_I}{\partial \psi} = \begin{bmatrix} -\cos\theta\sin\psi - \sin\varphi\sin\theta\cos\psi & \cos\theta\cos\psi - \sin\varphi\sin\theta\sin\psi & 0 \\ -\cos\varphi\cos\psi & -\cos\varphi\sin\psi & 0 \\ -\sin\theta\sin\psi + \sin\varphi\cos\theta\cos\psi & \sin\theta\cos\psi + \sin\theta\cos\theta\sin\psi & 0 \end{bmatrix}. \tag{2.84}$$

2.6.1.3 Simulation Analysis

The simulation is performed, taking the three-axis stabilized spacecraft as an example, with the singular-vector-based method and two-vector-based method under the same simulation conditions.

Assuming the spacecraft is of inertial space orientation, the attitudes in three directions stay the same, assumed to be $\varphi = 5°$, $\theta = 2°$, $\psi = 3°$. The simulation duration is 100 s, and the step is 0.5 s. Assume that the variance of process noise is $\boldsymbol{Q} = 10^{-8}\boldsymbol{I}_{3\times3}$ ($\boldsymbol{I}$ is the unit matrix) and covariance of measurement noise is $\boldsymbol{R} = 7.5 \times 10^{-7}\boldsymbol{I}_{3\times3} rad^2$ and that the covariance matrix of the initial estimate is $\boldsymbol{P}_{0/0} = \left(10^{-6}\boldsymbol{I}_{3\times3}\right) rad^2$. When the initial error of three attitudes ($\Delta\alpha_0$) are taken as 0.5°, 1.5° and 3° respectively, the filtering results are shown in Figs. 2.11, 2.12 and 2.13.

Table 2.3 shows the mean of attitude estimation errors, which is the statistical result of simulation data after filtering is steady.

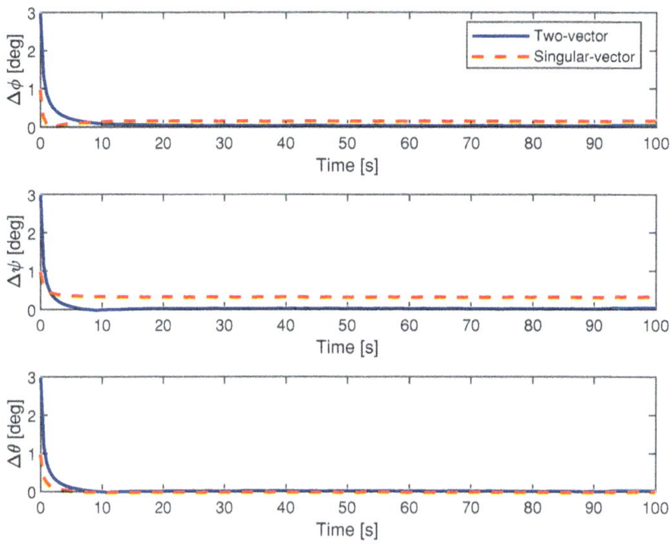

Fig. 2.11 Attitude filtering result when initial deviation is 0.5°

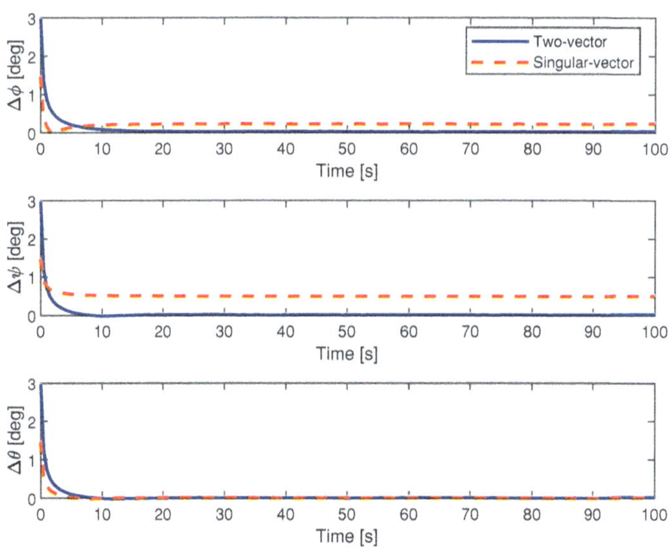

Fig. 2.12 Attitude filtering result when initial deviation is 1.5°

From the above results, under the same simulation conditions, the two-vector-based spacecraft attitude determination method is superior to the singular-vector-based method.

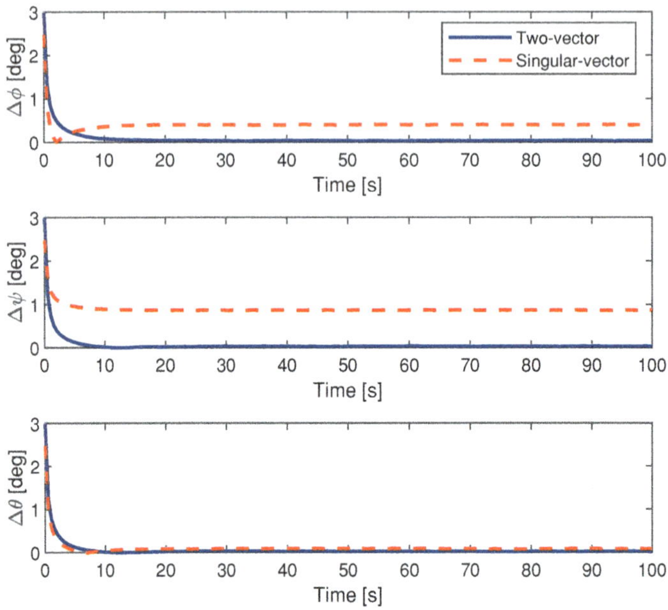

Fig. 2.13 Attitude filtering result when Initial deviation is 3°

Table 2.3 Statistical result of attitude filtering (mean error)

		$\Delta\alpha_0 = 0.5°$	$\Delta\alpha_0 = 1.5°$	$\Delta\alpha_0 = 3°$
Two-vector-based method	$\Delta\phi$	0.0370°	0.0396°	0.0393°
	$\Delta\theta$	0.0261°	0.0238°	0.0248°
	$\Delta\psi$	0.0296°	0.0295°	0.0294°
Singular-vector-based method	$\Delta\phi$	0.1837°	0.2525°	0.4075°
	$\Delta\theta$	0.0104°	0.0391°	0.1140°
	$\Delta\psi$	0.3522°	0.5168°	0.8748°

2.6.2 Means of Realizing Direction via the Observation of Pulsar

2.6.2.1 X-Ray Mapper

An X-ray mapper determines the attitude by imaging the sky through the coded aperture mask and pixelated detector, and then correlates the images with the X-ray sky map. This method is similar to the traditional star sensor. The difference is that X-rays are different from the ordinary visible light or ultraviolet rays, thus X-rays sources are imaged through the coded aperture mask and pixelated detector.

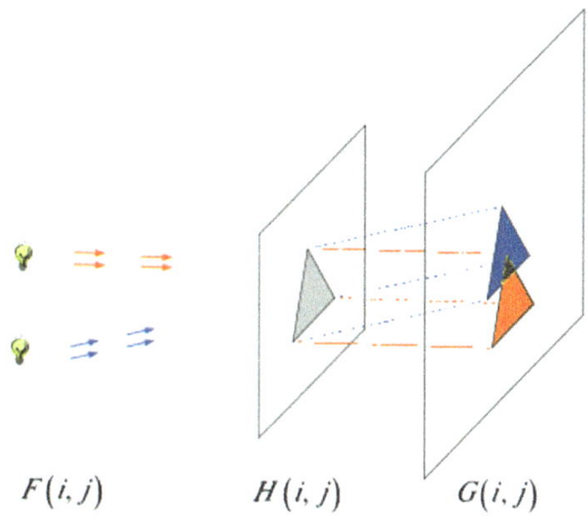

Fig. 2.14 Diagram of coded aperture imaging

Coded aperture imaging, which is developed from the pinhole imaging, includes the coding and decoding processes (Fig. 2.14). In the coding process, the coded aperture camera is used for collecting information about target sources as much as possible to obtain graphic overlaps of the target. The number of graphic overlaps is generally equal to that of apertures on the coding plate. The decoding process is used for reforming and processing graphic overlaps so as to regain a sharp image of the original target.

A graphic overlap of the target (X-ray source) is formed through the coded aperture camera. Then the obtained graphic overlap is processed with the optical or computer aided method to obtain a sharp image of the target. Assuming that the function of the target is $G(x, y)$, the image obtained through the coded aperture camera is

$$G(x, y) = F(x, y) \otimes H(x, y) + N(x, y) \tag{2.85}$$

where $\otimes$ is the convolution symbol, $H(x, y)$ is the convolution function of the coded aperture, $N(x, y)$ is the unrelated noise function, and $G(x, y)$ is the graphic overlap function.

The decoding process is a deconvolution process in fact. The convolution calculation, in the coding process, is equivalent to a low-pass filter for input signals, so the high-frequency component of input signals may be suppressed or lost. During decoding, to obtain high-resolution visible images of the target, the lost high-frequency component shall be regained, which will inevitably amplify high-frequency noise in the observed image. Therefore, it is necessary to take full account of weighing signal recovery with noise suppression based on the transfer function so as to gain the optimal recovery effect.

For coded aperture imaging, there is a large field of view and a point target after going through the coded aperture is effective for the whole receiving surface. So the coded aperture imaging technology reserves the high-resolution of the singular aperture imaging. Moreover, the image of the target source can be obtained even by using the weak X-rays radiation because the capability of collecting X-rays is increased by orders of magnitude. This technology is not very sensitive to noise interference, so it can obtain a high imaging SNR and light collecting efficiency.

Commonly, the characteristic parameters of the X-ray sources within the field of view are to match with that of the X-ray sources in the database. Once the pulsar is identified, the spacecraft attitude can be determined by the means listed in Sect. 2.6.1. The accuracy of this method is influenced by factors such as the area of the X-ray sensor, intensities of the X-ray source and background noise and duration of exposure. In the actual flight applications, since the navigation pulsars within the field of view cannot be selected randomly, the most effective way to shorten the duration of exposure is to increase the detector area.

2.6.2.2 X-Ray Scanner

An X-ray scanner, mainly consisting of an X-ray detector, a collimator and relevant electronic systems, is a collimated sensor system. X-rays travels along a straight line. The smaller the deviation in the directions of the X-ray source and the collimator's axis is, the stronger the signal received by the detector is. So the energy responses of the sensor should be fitted and the maximum response corresponds to the direction of the X-ray source.

For a spinning stabilization spacecraft, the sensor is fixed on the spacecraft, with the line of sight perpendicular to the spin axis of the spacecraft, and a main axis of the sensor is parallel with the spinning axis of the spacecraft. Then the X-ray sources can be scanned when the spacecraft is spinning and thus the spacecraft attitude can be measured. For a three-axis stabilized spacecraft, the sensor can be installed on a two-axis gimbal system which will enable the detector to scan.

For an X-ray scanner, if only one X-ray source is scanned every time, only the roll and pitch can be measured, but the yaw cannot. To cover this shortage, a differential collimator scanner can be used for observing the same X-ray source to read the whole spacecraft attitude information.

References

1. Qi G (2006) Fundamentals of time science. Higher Education Press, Peking
2. Xi X, Wang W, Gao Y (2003) Low earth space orbit foundation. National University of Defense and Technology Press, Changsha
3. Shuai P, Li M, Chen S, Huang Z (2009) Principle and method of X-ray pulser navigation system. China Aerospace Publishing House, Peking
4. Fei B. The application of relativity in modern naigtion. National Defence Industy Press, Peking
5. Irwin A, Fukushima T (1999) A numerical time ephemeris of the earth. Astron Astrophys 8(384):642–652
6. Lyne A, Graham-Smith F (1998) Pulsar astronomy. Cambridge University Press, Cambridge
7. Hobbs G, Lyne A, Kramer M (2010) An analysis of the timing irregularities for 366 pulsars. Mon Not R Astron Soc 402:1027–1048

8. Coles W, Hobbs G, Champion D et al (2011) Pulsar timing analysis in the presence of correlated noise. Mon Not R Astron Soc 418:561–570
9. Hongfei Ren (2012) Research on the pulsar navigation model under the framework of relativity theory. University of Information Engineering, Zhengzhou
10. Battin RH (1999) An introduction to the mathematics and methods of astrodynamics. AIAA
11. Jia P, Chen K, He L (1993) Long-range rocket ballisttics. National University of Defense Technology Press, Chagsha
12. Li J (2003) Orbit determination of spacecraft. National Defence Industy Press, Peking
13. Bar-Itzhack IY (1993) The program: an essay tool for angular position and rate computations. J Astronaut Sci 41(4):519–529
14. Jones DW (1995) Quaternions quickly transform coordinate without error buildup. EDN 40(5):95–100
15. BarItzhack IY, Idan M (1987) Recursive attitude determination from vector observations: euler angle estimation. J Guid Control Dyn 10(2):152–157
16. Dvornychenko VN (1985) The number of multiplications required to chain coordinate transformations. J Guid Control Dyn 8(1):157–159
17. Landis Markley F (2003) Attitude error representations for Kalman filtering. J Guidance Control Dyn 26(2):311–319
18. Phillips WF, Hailey CE, Gebert GA (2000) A review of attitude kinematics for aircraft simulation. AIAA modeling and simulation technologies conference and exhibit
19. Broucke RA (1993) On the use of poincare surfaces of section in rigid body motion. J Astronaut Sci 41(4):593–601

Chapter 3
X-Ray Pulsar Signal Processing

As X-ray pulsar signal processing is one of the key techniques of X-ray pulsar-based navigation system, this chapter focuses on investigating the way how to effectively and accurately extract the pulse TOA out of the photon TOAs recorded by the X-ray sensors.

3.1 X-Ray Pulsar Signal Model

The event of X-ray photon arriving at the spacecraft follows the on-homogeneous Poisson process, and the probability of M photons arriving for the interval of $\left[\, t_0 \ t_f \,\right]$ is given by

$$\Pr\left[K;\left(t_0\ t_f\right)\right]=\frac{\left(\int_{t_0}^{t_f}\lambda(t)dt\right)^M\exp\left(-\int_{t_0}^{t_f}\lambda(t)dt\right)}{M!} \tag{3.1}$$

where $\lambda(t) > 0$ is the photon rate function of pulsar signal, which can be given by

$$\lambda(t)=\alpha h(\phi_{\mathrm{det}}(t))+\beta \tag{3.2}$$

where

$$\alpha=\lambda_p A, \tag{3.3}$$

$$\beta=\lambda_n A. \tag{3.4}$$

In Eq. (3.2), α is the pulsar flow rate, β is the noise flow rate, $h(\cdot)$ is the normalized profile function of pulsar, and $\phi_{\mathrm{det}}(t)$ is the detected pulsar signal phase at t. In Eqs. (3.3) and (3.4), λ_p and λ_n are the mean flux density of the pulse signal and of the background, respectively, A is the sensor effective area.

W. Zheng and Y. Wang, *X-ray Pulsar-based Navigation*, Navigation: Science and Technology 5, https://doi.org/10.1007/978-981-15-3293-1_3

With the spacecraft motion considered, the phase propagation model of pulsar signal is given by

$$\phi_{\det}(t) = \phi_0 + f_s(t - t_0) + \frac{f_s}{c}\int_{t_0}^{t} v(\tau)d\tau \tag{3.5}$$

where ϕ_0 is the initial phase at t_0, f_s is the frequency of pulsar signal, c is the speed of light, and v is the projection of spacecraft velocity in the direction of pulsar.

3.2 Profile Recovery

3.2.1 *Epoch Folding*

Epoch folding is a method of recovering the pulse profile, which has been widely applied in astronomy and astrophysics.

By assuming the observation period to be T_{obs} comprising N_p pulsar periods, namely, $T_{obs} \approx N_p P$. The pulsar period P can be subdivided into N_b bins with each of T_b in length. Epoch folding can be realized as follows: (1) Fold the photon TOA sequence recorded in the subsequent period to the first period; (2) Calculate the photon number of each bin; and (3) Recover the profile by normalizing the photon number.

In the ith bin, the profile function $\tilde{\lambda}(T_i)$ for $i \in [1, N_b]$ can be given by

$$\tilde{\lambda}(T_i) = \frac{1}{N_p T_b}\sum_{j=1}^{N_p} c_j(T_i) \tag{3.6}$$

where $c_j(T_i)$ is the photon number folded into the ith bin at the jth period and T_i is the mid-time of the bin.

Performance of epoch folding depends on the accuracy of prior information of pulsar signal period. By taking the PSR B0531+21 for example, Fig. 3.1 illustrates the profiles obtained for different periods via epoch folding. The recovered profile may deform even for periodic error $= 10^{-4}\%$. For this reason, it is recommended to firstly determine the period or rotation frequency of pulsar signal in epoch folding, as detailed in Sect. 3.2.2.

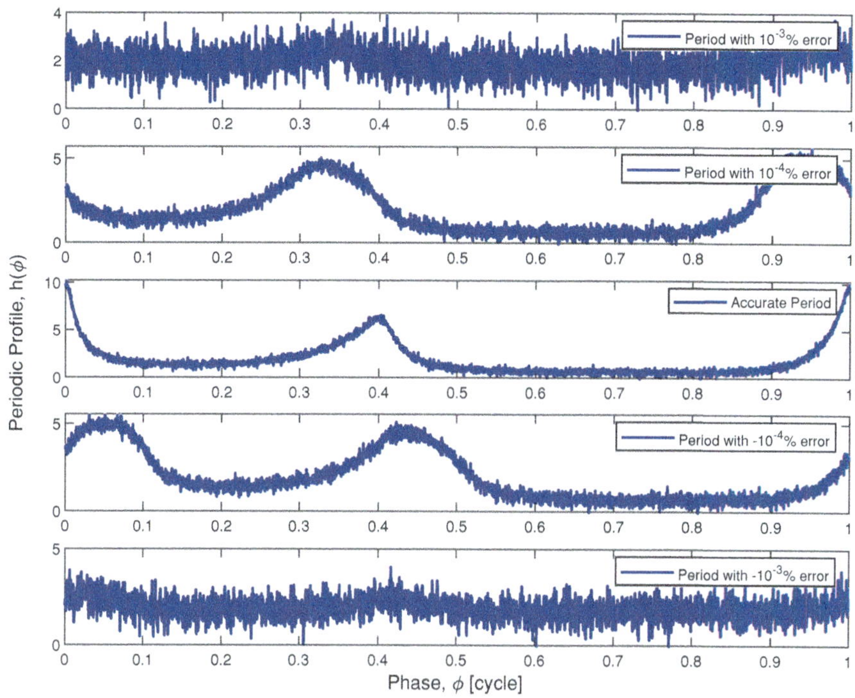

Fig. 3.1 Profiles for different periods by epoch folding

3.2.2 *Period Search*

The period of pulsar can be gained by time-domain search or frequency-domain analysis. Application of frequency-domain analysis requires recovering of light curve (non-epoch folding of photon TOA sequence). In order to have a high-precision period search result, the light curve should have a sampling step approaching to zero, which challenges the memory storage of computers. In contrast, the time-domain search methods all need to first recover the pulsar profile by epoch folding and thus are computationally efficient.

This subsection will establish a generalized framework for time-domain period search method and analyzing the performance.

3.2.2.1 Theoretical Derivation

As shown in Sect. 3.2.1, the signal period is used implicitly for recovering a pulsar profile. Thus, the recovered profile can be regarded as the implicit function $\tilde{\lambda}(P)$ of signal period. Optimization of signal period can be established by analyzing the statistical property of $\tilde{\lambda}(P)$

$$\begin{cases} \max S\left[\tilde{\lambda}(P)\right] \\ s.t.\ \ P \in \left[P_{\min}\ P_{\max}\right] \end{cases} \tag{3.7}$$

where $P_{\min}$ and $P_{\max}$ are the possible minimum and the maximum of signal period, respectively.

In this case, the key of signal period search lies in building an appropriate cost function for period search. Equations (3.8) and (3.9) provide $S\left[\tilde{\lambda}(P)\right]$ for searching methods of χ^2 and minimum entropy, respectively.

$$S\left[\tilde{\lambda}(P)\right] = \sum_{i=1}^{N_b} \frac{(n_i - N/N_b)^2}{N/N_b} \tag{3.8}$$

$$S\left[\tilde{\lambda}(P)\right] = \sum_{i=1}^{N_b} \tilde{p}(T_i)\ln(\tilde{p}(T_i)) \tag{3.9}$$

In Eq. (3.8), $n_i = \sum_{j=1}^{N_p} c_j(T_i)$ is the number of photons in the ith bin, $N = \sum_{j=1}^{N_p}\sum_{i=1}^{N_b} c_j(T_i)$ is the number of all photons within the observation period. In Eq. (3.9), $\tilde{p}(T_i) = n_i/N$.

Based on the above generalized framework of period search, we could derive a period search method based on likelihood function.

By changing the regularization factor in Eq. (3.6), the recovered profile can be regarded as a histogram with multivariate independent distribution so as to obtain the corresponding probability function given by [1]

$$P\left(T_1 \cdots T_i \cdots T_{N_b} \left| P\right.\right) = \frac{N!}{\prod_{i=1}^{N_b} n_i!}\prod_{i=1}^{N_b}\tilde{p}(T_i). \tag{3.10}$$

Then, the logarithmic form of Eq. (3.10) can be expressed as

$$LLF\left(T_1 \cdots T_i \cdots T_{N_b} \left| P\right.\right) = C + \sum_{i=1}^{N_b} n_i \ln \tilde{p}(T_i) \tag{3.11}$$

where $C = \ln N! - \sum_{i=1}^{N_b} \ln n_i!$ is irrelevant to $\tilde{p}(T_i)$ and can be removed. Therefore, $S\left[\tilde{\lambda}(P)\right]$ for likelihood function can be modified as

$$S\left[\tilde{\lambda}(P)\right] = \sum_{i=1}^{N_b} n_i \ln \tilde{p}(T_i). \tag{3.12}$$

Cost functions shown in Eqs. (3.8), (3.9) and (3.12) are convex functions within the taking value interval, so there exist maximum values in the potential interval. As stated in Ref. [2], the performance of time-domain search method is little affected by the detailed expression of cost function. Thus, we analyze the performance of period search by just taking the likelihood function for example.

3.2.2.2 Performance Analysis

The performance of time-domain period search method is verified by simulated and real data.

(1) **Application for simulation data**

PSR B0531+21 is taken as the simulation object, with the observation period assumed to be 3000 s. Table 3.1 lists its simulation parameters and Fig. 3.2 shows its simulated profile. The spacecraft is assumed to be stationary, and the photon TOAs are simulated via the use of method in [3].

Table 3.1 Simulation parameters of PSR B0531+21

Period [ms]	33.4
α [ph/s]	660
β [ph/s]	13860.2
Initial phase [cycle]	0.3

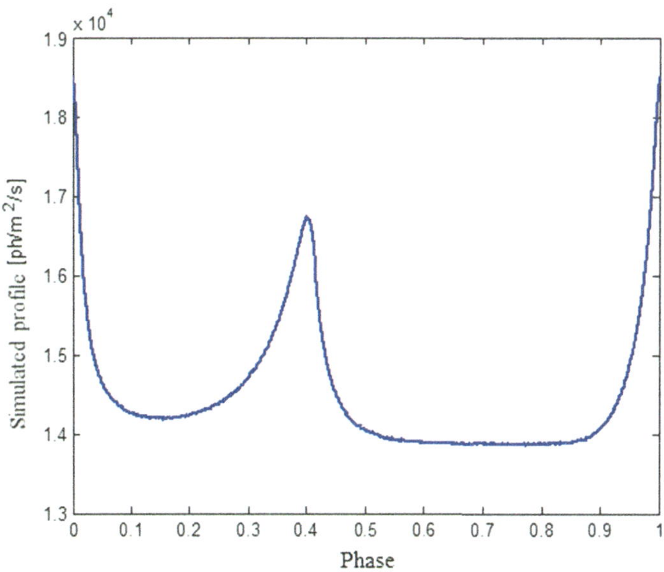

Fig. 3.2 Simulated profile of PSR B0531+21

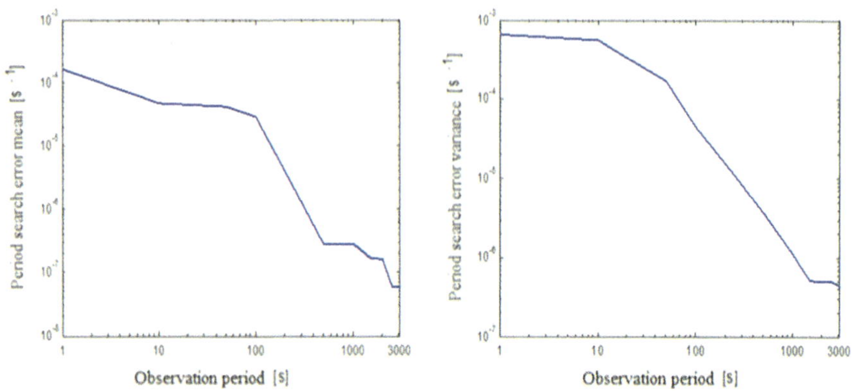

(a) Period search error mean (b) Period search error variance

Fig. 3.3 Period search result of PSR B0531+21

Figure 3.3 shows the period search error with respect to the observation period. As the observation period keeps increases, the period search error would rapidly reduce when the observation period is less than 1000 s but degrade at a slower speed when the observation period overtakes 1000 s.

(2) **Application for real data**

Feasibility of the method is further verified through the measured data of RXTE satellite. The PSR B0531+21 data package with ID of 20804-01-01-00 is selected. The software Heasoft is used to filter the good time interval and to transfer the photon TOAs recorded by the satellite to those at SSB. The observation period for the data observed is chosen to be 1000 s. It should be noted that flux for processing of real data is measured in photon/s (ph/s). The sensor effective area is related to the selected energy range. As there is not an analysis on the pulsar energy spectrum in this book, the flux of pulsar will not be further normalized by the effective area of sensor.

Figure 3.4 shows the likelihood function with respect to the search frequency in which the likelihood function arrives at the maximum when the search frequency becomes 29.8790027 s^{-1}. It means the appropriate period for the selected data is 33.4683 ms. We recover the pulsar profile by using period of 33.4683 ms and of 33.4 ms respectively, as shown in Figs. 3.5 and 3.6. The former period can provide stable pulse profile, while the latter cannot. Therefore, period search is necessary for processing the real pulsar data.

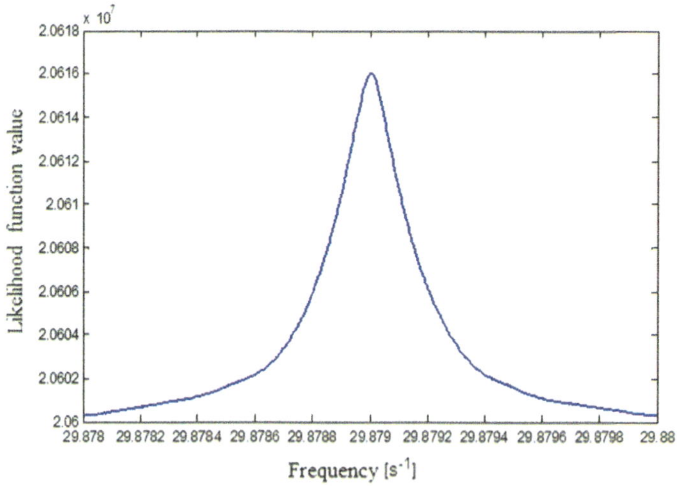

Fig. 3.4 Likelihood function versus frequency

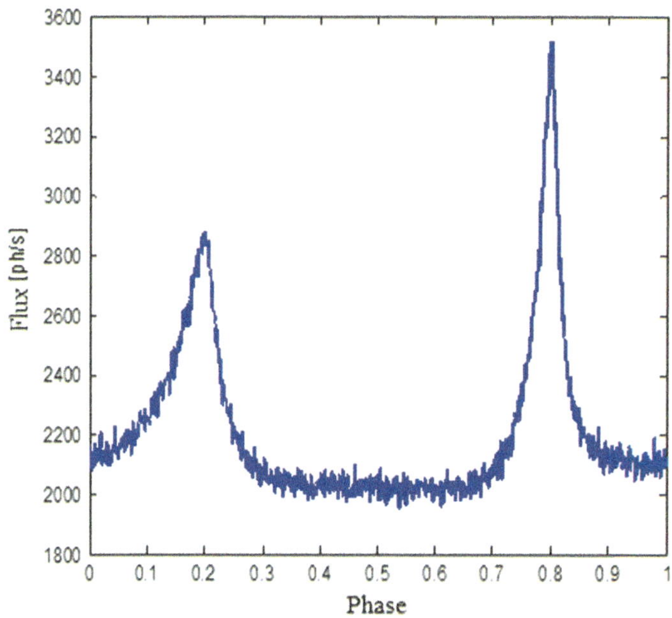

Fig. 3.5 Profile recovered using searched period

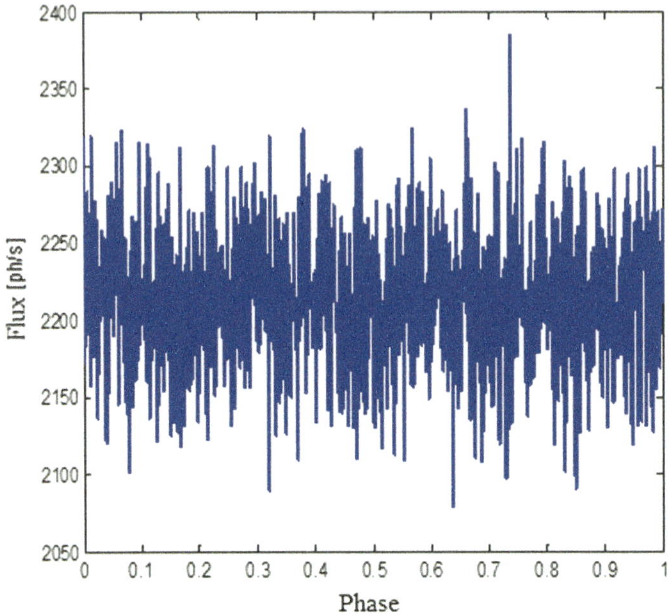

Fig. 3.6 Profile folded for 33.4 ms

3.2.3 *Enhancing the Signal to Noise Ratio of Profile*

When the epoch folding and period search have been done, an empirical profile can be recovered. This subsection will investigate how to enhance the signal to noise ratio (SNR) of the empirical profile.

3.2.3.1 Profile Recovering via Searching for the Optimum N_b

As stated in Ref. [3], the empirical profile can be viewed as a profile with high SNR (referred to be the target profile in the remainder of the subsection) contaminated by a zero-mean Gaussian white noise, i.e.,

$$\tilde{\boldsymbol{\lambda}} = \bar{\boldsymbol{\lambda}} + \boldsymbol{w}(\phi_0) \tag{3.13}$$

where $\tilde{\boldsymbol{\lambda}}$ and $\bar{\boldsymbol{\lambda}}$ are the empirical profile and the target profile, respectively, which can be expressed as

$$\tilde{\boldsymbol{\lambda}} = \left[\, \tilde{\lambda}(T_1)\ \tilde{\lambda}(T_2)\ \cdots\ \tilde{\lambda}(T_{N_b}) \,\right]^T \tag{3.14}$$

$$\bar{\boldsymbol{\lambda}} = \left[\, \bar{\lambda}(T_1)\ \bar{\lambda}(T_2)\ \cdots\ \bar{\lambda}(T_{N_b}) \,\right]^T. \tag{3.15}$$

In Eq. (3.13), $\boldsymbol{w}(\phi_0)$ is the epoch-folded noise which can be molded as a zero-mean Gaussian white noise with a variance given by

$$\text{var}[\boldsymbol{w}(\phi_0)] = \frac{N_b}{T_{obs}} diag[\bar{\boldsymbol{\lambda}}(\phi_0)]. \tag{3.16}$$

From Eq. (3.16), any increase of N_b will increase the variance of epoch-folded noise and thereby results in great fluctuation of empirical profile, namely, over-fitting. However, a smaller N_b will lead to the sub-fitting of empirical profile, resulting in an extremely smooth profile which cannot reflect the true details of the profile. To make best use of observed data, it is quite essential to find the optimum N_b.

(1) **Theoretical derivation**

In this case, the empirical profile may be regarded as an implicit function of N_b. Therefore, the optimization problem for N_b can be established as

$$\begin{cases} \min \ S_2\left[\tilde{\boldsymbol{\lambda}}(N_b)\right] \\ s.t. \ \ N_b \in \left[N_{b,\min} \ N_{b,\max}\right] \end{cases}. \tag{3.17}$$

The key to Eq. (3.17) become the proposal of an appropriate expression of $S_2\left[\tilde{\boldsymbol{\lambda}}(N_b)\right]$.

We use the integral square error (ISE) to derive $S_2\left[\tilde{\boldsymbol{\lambda}}(N_b)\right]$. The ISE can be expressed as

$$ISE = \sum_{i=1}^{N_b} \left(\tilde{\lambda}(T_i) - \bar{\lambda}(T_i)\right)^2 \tag{3.18}$$

where $\bar{\lambda}(T_i)$ is the target profile.

Equation (3.18) can be expanded as

$$ISE = \sum_{i=1}^{N_b} \tilde{\lambda}(T_i)^2 - 2\sum_{i=1}^{N_b} \tilde{\lambda}(T_i)\bar{\lambda}(T_i) + \sum_{i=1}^{N_b} \bar{\lambda}(T_i)^2 \tag{3.19}$$

where $\sum_{i=1}^{N_b} \bar{\lambda}(T_i)^2$ is irrelevant to the recovered profile, with $\sum_{i=1}^{N_b} \tilde{\lambda}(T_i)\bar{\lambda}(T_i) \approx E\left[\tilde{\lambda}(T_i)\right]$.

$E\left[\tilde{\lambda}(T_i)\right]$ is too complicate to calculate directly, so the cross-validation method [1, 4, 5] is adopted to estimate $E\left[\tilde{\lambda}((T_i))\right]$. For a profile with N_b bins, we sequentially remove data in bin of T_i, and use the remainder data to predict the value of $\tilde{\lambda}(\cdot)$ at T_i,

$\tilde{\lambda}_{-1}(T_i)$. Repeating N times and taking the average of all estimates yield the estimate of $E\left[\tilde{\lambda}(T_i)\right]$. Rudemo has demonstrated that it is unbiased estimate of $E\left[\tilde{\lambda}(T_i)\right]$.

Therefore, assuming $S_2\left[\tilde{\boldsymbol{\lambda}}(N_b)\right] = \sum_{i=1}^{N_b} \tilde{\lambda}(T_i)^2 - 2\sum_{i=1}^{N_b} \tilde{\lambda}(T_i)\bar{\lambda}(T_i)$ yields

$$S_2\left[\tilde{\boldsymbol{\lambda}}(N_b)\right] = \sum_{i=1}^{N_b} \frac{n_i^2}{N^2 T_b^2} - \frac{2}{N}\sum_{i=1}^{N_b} \tilde{\lambda}_{-1}(T_i). \tag{3.20}$$

Simplifying Eq. (3.20) yields

$$S_2\left[\tilde{\lambda}(N_b)\right] = \frac{2N_b}{(N-1)P} - \frac{(N+1)N_b}{N^2(N-1)P}\sum_{i=1}^{N_b} n_i^2. \tag{3.21}$$

(2) **Performance analysis**

Feasibility of the method is verified by using the measured data of RXTE satellite given in Sect. 3.2.2.2. Figure 3.7 shows the variation of S_2 over N_b, based on which, the optimum N_b is 250. Figures 3.8, 3.9 and 3.10 depict the profiles recovered with N_b equaling to 250, 50, and 1000, respectively. By comparison with Figs. 3.9 and 3.10, the profile in Fig. 3.8 which is recovered by using the optimum N_b has retainable details and high SNR.

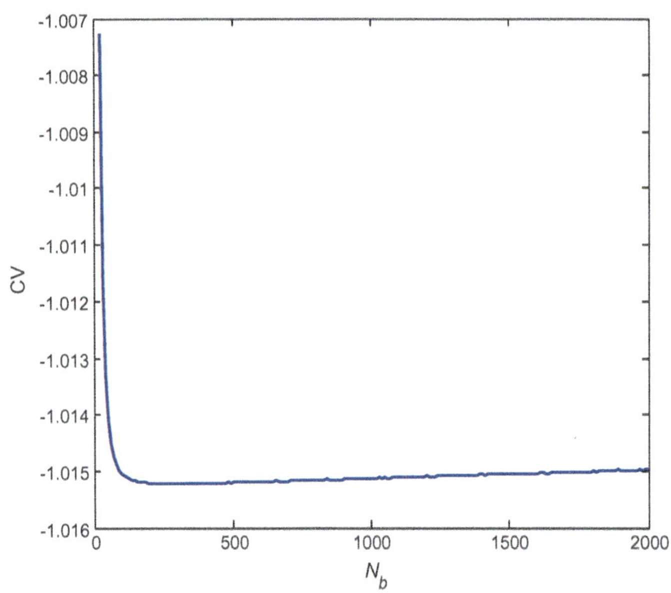

Fig. 3.7 CV versus N_b

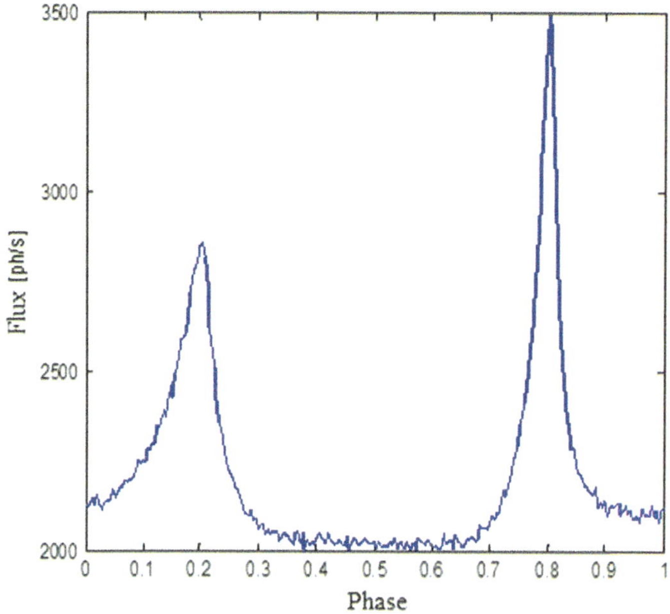

Fig. 3.8 Recovered profile ($N_b = 250$)

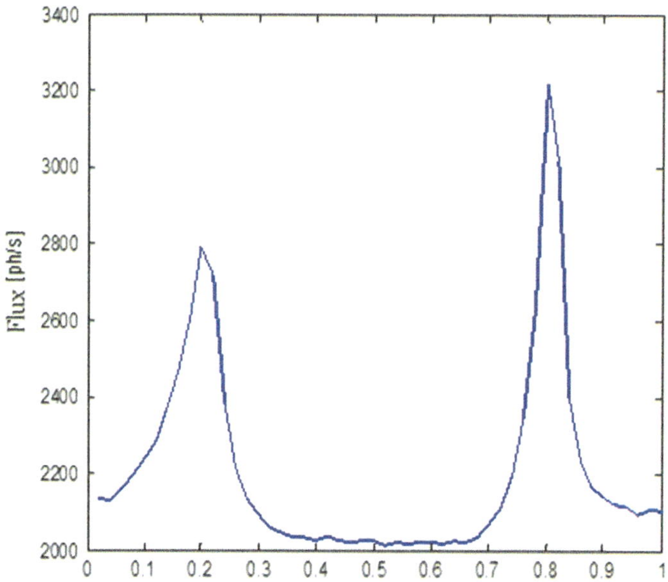

Fig. 3.9 Recovered profile ($N_b = 50$)

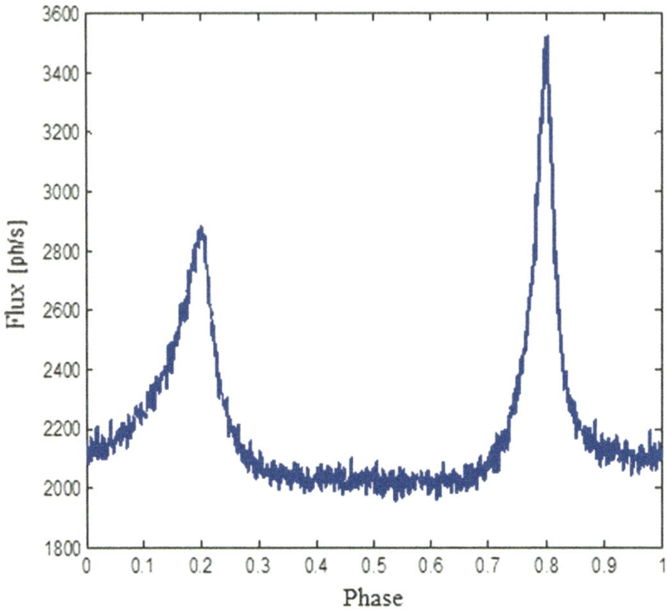

Fig. 3.10 Recovered profile ($N_b = 1000$)

It should be noted that for high-resolution (Bin with small width) profile, the analytic expression can be obtained by means of Gaussian function fitting after the profile is recovered using the optimum N_b.

3.2.3.2 Profile De-noising Based on Kernel Regression

Kernel regression is a non-parametric approach which is less restrictive to distribution of observation and more adaptive than parametric regression. It has been successfully applied to image de-noising and machine learning with good results [4]. The type of kernel function, although needed in kernel regression, has little effect on the final regression result [4]. The key of kernel regression is selection of kernel bandwidth, and Leave One Out Cross Validation (LOOCV) is commonly used for this purpose [5].

(1) **Related theories**

1) **Kernel regression method**
Assuming a non-linear model is

$$y = s(X) + \omega \tag{3.22}$$

where $\boldsymbol{y} \in \mathbb{R}^{n\times 1}$ is the measurement, $\boldsymbol{X} \in \mathbb{R}^{n\times 1}$ is the design point, and $\boldsymbol{\omega}$ is the measurement noise.

The estimate of $\boldsymbol{y}$ at x can be expressed as

$$\hat{y}(x) = \sum_{i=1}^{n} W_i(x) y_i \tag{3.23}$$

where y_i is the measurement ith component of $\boldsymbol{y}$, and $W_i(x)$ is the weight function with the expression of

$$W_i(x; X_1, \cdots, X_n) = \kappa\left(\frac{x - X_i}{h}\right) / \sum_{i=1}^{n} \kappa\left(\frac{x - X_i}{h}\right). \tag{3.24}$$

In Eq. (3.24), $\kappa(\cdot)$ is the kernel function and h is the kernel bandwidth. Gaussian kernel function as shown in Eq. (3.25) is commonly used.

$$\kappa(t) = \frac{1}{\sqrt{2\pi} h} \exp\left(-\frac{t^2}{2h^2}\right) \tag{3.25}$$

As shown in Eq. (3.23), this non-parametric method has no limitation on the regression model and therefore has a wide scope of application. However, the regression bandwidth h still needs to be determined for non-parametric model.

2) **Cross-validation method**

In kernel regression, the bandwidth h is used for data smoothing. A too wide h will result in an excessive smooth whereas a narrow h provides limited smoothing effect. Clark proposed the cross-validation method for the selection of h [5] by wiping a set of data (X_i, y_i) to obtain the predicted value from remaining n-1 components and by calculating the difference between the predicted value and the actual observed value. The optimum kernel bandwidth h should ensure a minimum difference, that is, h is given by:

$$h = \arg\min\left(\sum_{i=1}^{n} \left(\hat{y}^{-i} - y_i\right)^2\right) \tag{3.26}$$

where $\hat{y}^{-i}$ is the predicted value obtained at the ith design point after the ith set of data is eliminated.

(2) **Performance analysis**

To verify the performance of proposed method, this section performs an analysis using the simulated data and the real data of RXTE respectively.

Table 3.2 Simulation parameters of PSR B1821−24

Period [ms]	3.05
α [ph/s]	1.93
β [ph/s]	50
Initial phase [cycle]	0.241

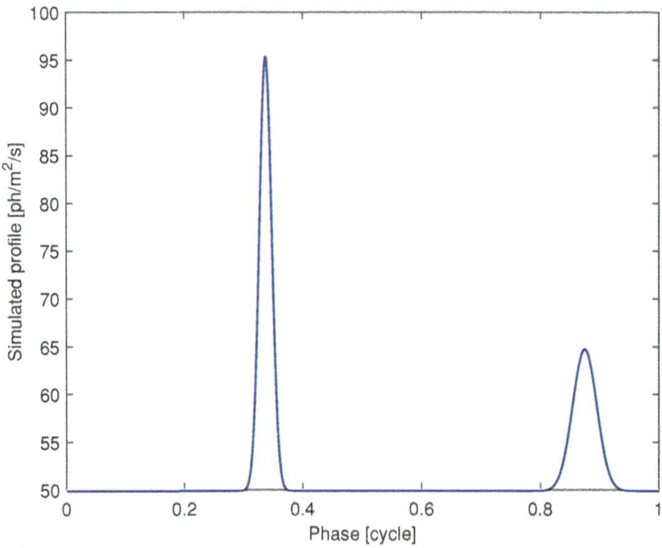

Fig. 3.11 Simulated profile of PSR B1821−24

1) **Simulation experiment analysis**
Data of two X-ray pulsars (PSR B0531+21 and PSR B1821−24) are simulated in this section. PSR B0531+21 is a young pulsar with a high flux, has an observation period of 1000 s and $N_b = 10000$, and PSR B1821−24 is a millisecond pulsar with low flux, has observation period of 3000 s and $N_b = 1000$. See Table 3.1 and Fig. 3.3 for simulation parameters and profile of PSR B0531+21, respectively, and Table 3.2 and Fig. 3.11 for PSR B1821−24.
The performances of wavelet denoising with soft threshold (WDST), empirical model decomposition (EMD), direct epoch folding (DEF) and kernel regression (KR) are analyzed and compared from two aspects of SNR and profile fidelity which is measured by two indices:
(a) Relative error of denoised profile, which is defined by

$$e_r = \sqrt{\sum_{i=1}^{N_b} \left(\tilde{\lambda}(T_i) - \bar{\lambda}(T_i)\right)^2 / \sum_{i=1}^{N_b} \bar{\lambda}^2(T_i)} \times 100\%. \tag{3.27}$$

(b) Phase deviation of denoised profile $\Delta\phi$. It is the deviation of denoised profile from the target profile and can be obtained by cross-correlation between the denoised profile and target profile. The phase deviation here is the result of 1000 Monte Carlo simulations.

Figures 3.12 and 3.13 show the profiles recovered by means of DEF, and profiles denoised by KR, WDST and EMD. From both figures, KR, WDST and EMD can reduce the impact of noise compared to DEF. Besides, WDST and EMD produce more glitches than KR does. Especially for signals of PSR B0531+21, WDST provides poor effect at the boundary and produces recovered profile with large distortion.

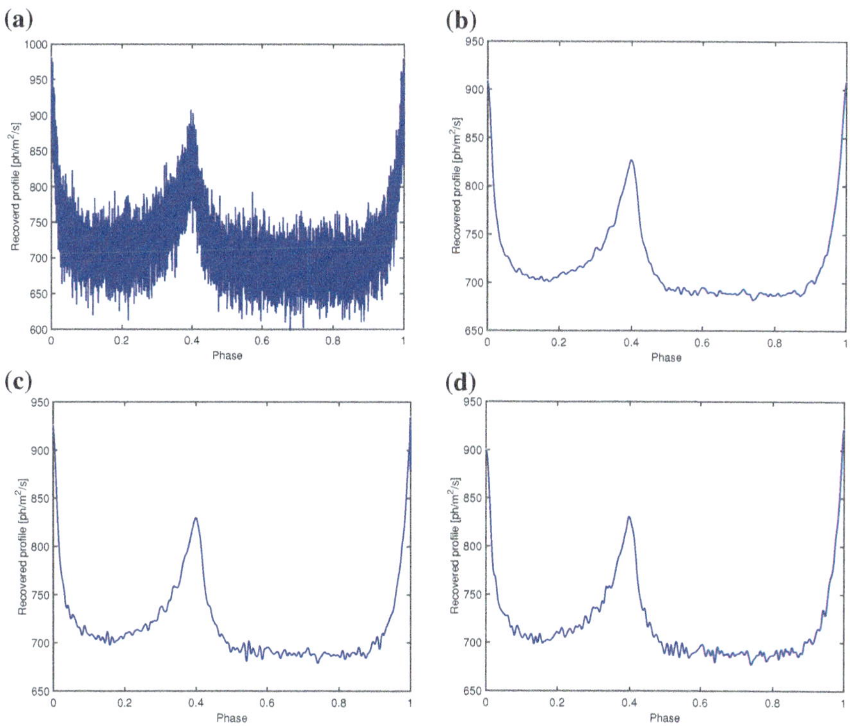

Fig. 3.12 Comparison between profile de-noising effects of PSR B0531+21

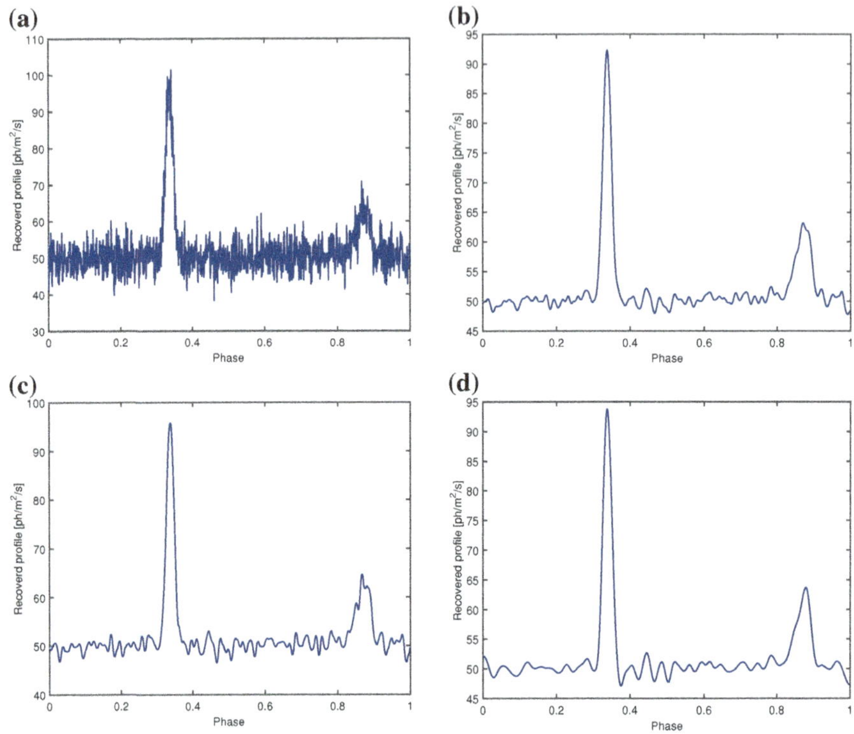

Fig. 3.13 Comparison Between Profile De-noising Effects of PSR B1821−24

Table 3.3 gives SNR, e_r and $\Delta\phi$ obtained by using these four methods and shows that KR, WDST and EMD can improve SNR of recovered profile and reduce e_r compared to DEF, while KR can provide the maximum SNR

Table 3.3 Comparison of four methods in performance

	SNR [dB]		er [%]		$\Delta\varphi$ [cycle]	
	PSR B0531+21	PSR B1821−24	PSR B0531+21	PSR B1821−24	PSR B0531+21	PSR B1821−24
KR	20.97	17.01	0.8	1.99	4.4×10^{-5}	1.3×10^{-3}
WDST	20.7	16.3	0.85	2.34	5.86×10^{-5}	1.36×10^{-3}
EMD	20.53	15.98	0.88	2.52	5.96×10^{-5}	1.43×10^{-3}
DEF	14.28	11.38	3.73	7.28	1.26×10^{-5}	1.8×10^{-3}

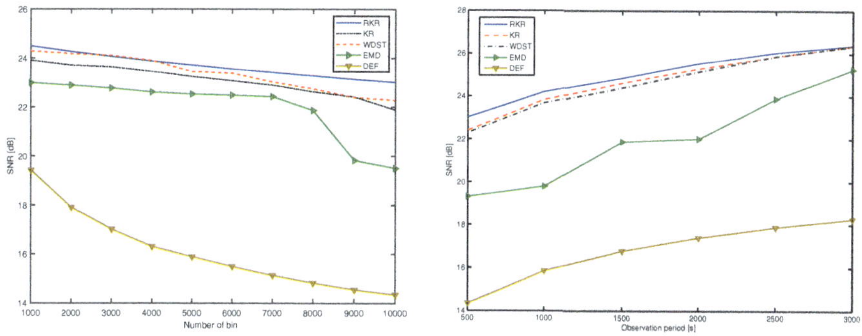

Fig. 3.14 Analysis of influencing factors of SNR

and the minimum e_r and $\Delta\phi$. This means KR has the best performance in the given simulation environment.

The relationship between the denoising performance versus N_b and observation period is analyzed below by taking PSR B0531+21 for example. Figure 3.14 shows SNR versus N_b and observation period for these methods. As N_b increases, SNR of profile decreases gradually. KR, WDST and EMD are at a speed slower than DEF. When N_b decreases, WDST behaves similar to KR, however the latter shows a gradually obvious superiority as N_b increases.

As the observation period grows, SNR increases, and KR and WDST have overlapping curves of SNR, which suggests that the increase of observation period will improve the performance of de-noising method. All de-noising methods will ultimately reach their limit of accuracy which depends on the shape of pulsar profile and the flux of pulsar. DEF does improve performance but to a limited extent which is still below that of the other three methods.

2) **RXTE satellite data analysis**

Figure 3.15 shows profiles recovered by using DEF, KR, WDST, EMD based on the same set of data of RXTE as given in Sect. 3.2.2.2. KR, WDST and EMD provide profiles with SNR higher than DEF, and KR produces fewer glitches, which accords with the aforementional simulation result.

A comparison with the result given in Sect. 3.2.3.1 suggests that empirical profiles with higher SNR can be obtained by de-noising or taking the optimum N_b. If sufficient photons are available within the observation period, de-noising is recommended to ensure details of profiles, otherwise, the latter method is recommended.

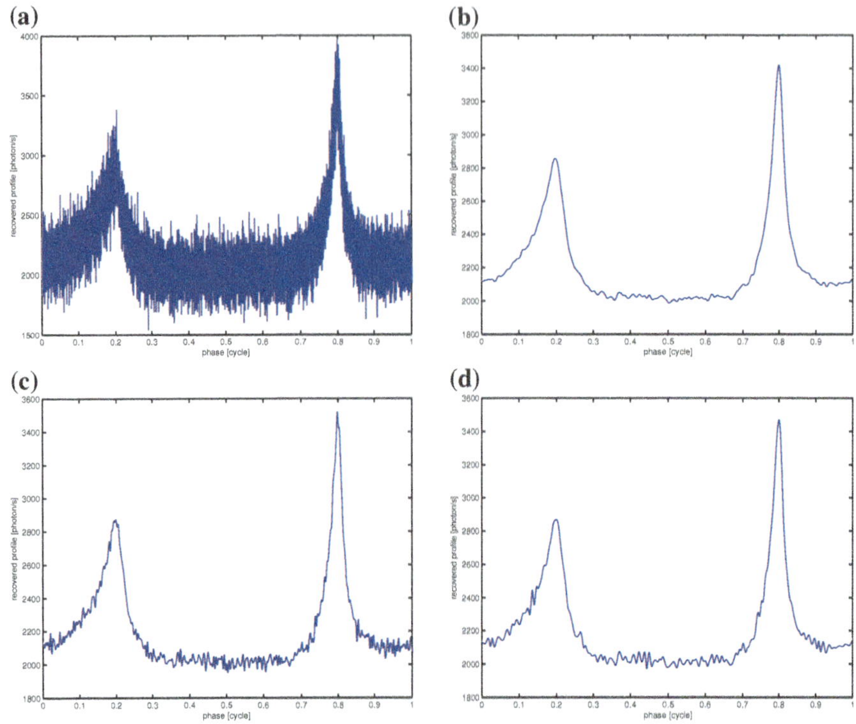

Fig. 3.15 De-noising result analysis based on RXTE data

3.3 Pulse TOA Calculation for Stationary Case

Based on the study in Sects. 3.1 and 3.2, this section will illustrate how to extract the pulse TOA out of the recorded photon TOAs in the case where the investigated spacecraft is assumed to stationary.

3.3.1 Pulse TOA Calculation Methods

3.3.1.1 Pulse TOA Calculation Method Based on Epoch Folding

After profile recovery, the initial phase of pulsar signal can be obtained by comparison with the template. This section generalizes the proposed methods for estimating the pulse TOA as a one-dimensional linear search problem, through which the global optimum of a cost function appropriately designed can be found out. The phase corresponding to the global optimum is the estimate of the initial phase.

In time-domain methods, the most representative ones include the cross correlation (CC), nonlinear least square (NLS) and fast near-maximum likelihood estimator (FNMLE). Their cost functions are given below respectively

$$J_{CC} = \sum_{i=1}^{N_b} \tilde{\lambda}(T_i)\lambda(T_i) \tag{3.28}$$

$$J_{NLS} = \sum_{i=1}^{N_b} \left(\tilde{\lambda}(T_i) - \lambda(T_i)\right)^2 \tag{3.29}$$

$$J_{FNMLE} = \sum_{i=1}^{N_b} \tilde{\lambda}(T_i) \log(\lambda(T_i)) \tag{3.30}$$

where $\lambda(T_i)$ is the template at T_i.

References [3, 6] discuss their performances separately which are similar in terms of the estimation accuracy of initial phase. But CC and FNMLE permit the use of fast Fourier transform (FFT) and can greatly save the computational cost. Therefore, it is applicable to a practical on-board computation.

Taylor method is the most famous frequency-domain method [1, 3, 4], and also is the default approach for pulse TOA calculation in the software PSRCHIVE. Its estimation result would not be affected by the bin length but greatly affected by the SNR of profile, costing a large calculation amount.

3.3.1.2 Pulse TOA Calculation Method via the Direct Use of Photon TOAs

Equation (3.5) can be simplified into

$$\phi_{\text{det}}(t) = \phi_0 + f_s(t - t_0). \tag{3.31}$$

According to the properties of Poisson distribution, by setting the photon TOA sequence as $\{t_i\}_{i=1}^M$, the joint probability density function of photon TOA sequence is given by [1–3]

$$p(\{t_i\}_{i=1}^M, M) = \begin{cases} e^{-\Lambda} \prod_{i=1}^{M} \lambda(\phi_{\text{det}}(t_i)) & M \geq 1 \\ e^{-\Lambda} & M = 0 \end{cases} \tag{3.32}$$

where

$$\Lambda \triangleq \Lambda(t_f) - \Lambda(t_0) \tag{3.33}$$

$$\Lambda(t) = \int_0^t \lambda(\xi)d\xi. \tag{3.34}$$

By defining Eq. (3.32) as the likelihood function, the log-likelihood function can be given by

$$LLF = \sum_{i=1}^{M} \ln(\phi_{\det}(t_i)) - \Lambda(\phi_0). \tag{3.35}$$

Reference [3] has demonstrated that the second term of Eq. (3.35) is a constant. So Eq. (3.35) can be transformed into

$$LLF = \sum_{i=1}^{M} \ln(\phi_{\det}(t_i)). \tag{3.36}$$

The initial phase of pulsar signal can be obtained by searching the maximum of Eq. (3.36). With the frequency of pulsar signal unknown, the estimates of initial phase and frequency can be obtained through two-dimensional searching by transforming Eq. (3.36) into the two-dimensional function of initial phase and frequency.

3.3.2 Performance Analysis

Suppose the background noise flux of PSR B0531+21 is 10 times of its pulse part flux and the background noise flux of PSR B1937+21 is 100 times of its pulse part flux. The detector area is 1 m^2. The width of bin is 1 μs. By using the cross-correlation method and performing 200 Monte Carlo simulations, the TOA estimation results are shown in Fig. 3.16. The TOA estimation accuracy could increase as the observation period increases. Because the flux of PSR B1937+21 is weaker than PSR B0531+21, TOA estimation accuracy for PSR B1937+21 is lower than that of PSR B0531+21 when the observation periods for PSR B1937+21 is as the same as for PSR B0531+21.

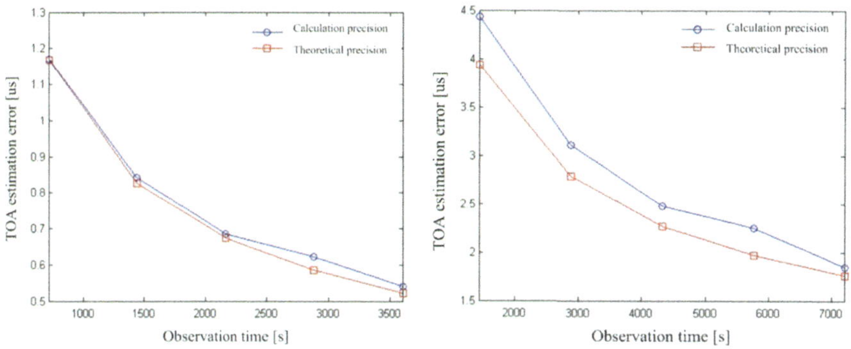

(a) PSR B0531+21 (b) PSR B1937+21

Fig. 3.16 Curve for the precision of the TOA to observation time

3.4 Pulse TOA Calculation for Dynamics Case

This section will focus on the pulse TOA calculation in the case where the investigated spacecraft is performing an orbital motion.

3.4.1 Improved Phase Propagation Model

In the solar system barycenter inertial system, configurations of the spacecraft and the pulsar are shown in Fig. 3.17.

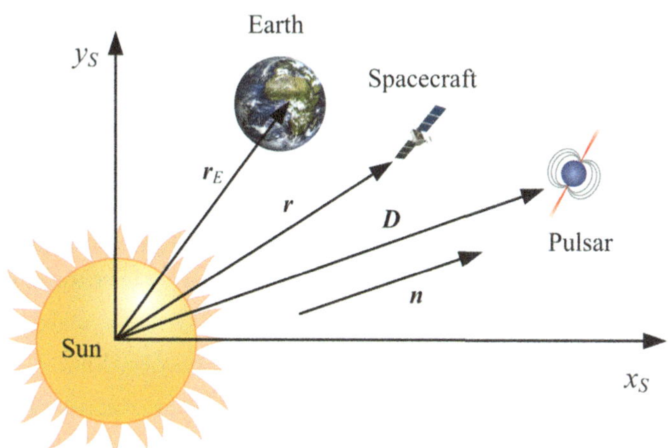

Fig. 3.17 Geometric construction of spacecraft and pulsar

Considering the fact that pulsar is far from the solar system, the distance between the pulsar and the spacecraft can be given as

$$d = |\boldsymbol{r} - \boldsymbol{D}| \approx \boldsymbol{n} \cdot (\boldsymbol{r} - \boldsymbol{D}). \tag{3.37}$$

Therefore, the velocity of spacecraft projected on the pulsar is expressed as

$$v = \boldsymbol{n} \cdot (\dot{\boldsymbol{r}} - \dot{\boldsymbol{D}}). \tag{3.38}$$

Substituting Eq. (3.38) into Eq. (3.5) yields

$$\phi_{\text{det}}(t) = \phi_0 + f_s(t - t_0) + \frac{f_s}{c} \int_{t_0}^{t} \boldsymbol{n} \cdot \left(\dot{\boldsymbol{r}}(\tau) - \dot{\boldsymbol{D}}(\tau)\right) d\tau. \tag{3.39}$$

Equation (3.39) can be transformed into

$$\phi_{\text{det}}(t) = \phi_0 + f_s(t - t_0) + \frac{f_s}{c} \int_{t_0}^{t} \boldsymbol{n} \cdot \dot{\boldsymbol{r}}(\tau) d\tau - \frac{f_s}{c} \int_{t_0}^{t} \boldsymbol{n} \cdot \dot{\boldsymbol{D}}(\tau) d\tau. \tag{3.40}$$

The solution of Eq. (3.40) is given as

$$\phi_{\text{det}}(t) = \phi_0 + f_s(t - t_0) + \frac{f_s}{c} \boldsymbol{n} \cdot (\boldsymbol{r}(t) - \boldsymbol{r}(t_0)) - \frac{f_s}{c} \boldsymbol{n} \cdot (\boldsymbol{D}(t) - \boldsymbol{D}(t_0)). \tag{3.41}$$

Considering that proper motion of pulsar during the observation period of spacecraft is negligible, $\boldsymbol{D}(t) \approx \boldsymbol{D}(t_0)$, then Eq. (3.41) can be transformed into

$$\phi_{\text{det}}(t) = \phi_0 + f_s(t - t_0) + \frac{f_s}{c} \boldsymbol{n} \cdot (\boldsymbol{r}(t) - \boldsymbol{r}(t_0)). \tag{3.42}$$

For near-earth orbit spacecraft, $\boldsymbol{r}$ can be given as

$$\boldsymbol{r} = \boldsymbol{r}_E + \boldsymbol{r}_{SC/E} \tag{3.43}$$

where $\boldsymbol{r}_E$ is the position vector of the earth relative to SSB and $\boldsymbol{r}_{SC/E}$ is the position vector of the spacecraft relative to the earth. Then Eq. (3.42) can be transformed into

$$\begin{aligned} \phi_{\text{det}}(t) = \phi_0 + f_s(t - t_0) + \frac{f_s}{c} \mathbf{n} \cdot \left(\mathbf{r}_{SC/E}(t) - \mathbf{r}_{SC/E}(t_0)\right) \\ + \frac{f_s}{c} \mathbf{n} \cdot (\mathbf{r}_E(t) - \mathbf{r}_E(t_0)). \end{aligned} \tag{3.44}$$

For deep space explorers, Eq. (3.42) can be directly used.

3.4.2 *Linearized Phase Propagation Model*

3.4.2.1 General Case

Based on orbital dynamics of spacecraft, the position of the spacecraft at any time can be represented by the estimated position and the corresponding orbital error, namely

$$\boldsymbol{r}(t) = \tilde{\boldsymbol{r}}(t) + \delta \boldsymbol{r}(t) \tag{3.45}$$

where $\tilde{\boldsymbol{r}}(\cdot)$ is the estimated position and $\delta \boldsymbol{r}(\cdot)$ is the error within $\tilde{\boldsymbol{r}}(\cdot)$. Then Eq. (3.42) can be transformed into

$$\phi_{\text{det}}(t) = \phi_0 + f_s(t - t_0) + \frac{f_s}{c}\boldsymbol{n} \cdot (\tilde{\boldsymbol{r}}(t) - \tilde{\boldsymbol{r}}(t_0)) + \frac{f_s}{c}\boldsymbol{n} \cdot (\delta \boldsymbol{r}(t) - \delta \boldsymbol{r}(t_0)). \tag{3.46}$$

$\delta \boldsymbol{r}(t)$ can be obtained by propagation of $\delta \boldsymbol{r}(t_0)$ and $\delta \boldsymbol{v}(t_0)$, namely

$$\delta \boldsymbol{r}(t) = \boldsymbol{\Phi}_{rr}(t, t_0)\delta \boldsymbol{r}(t_0) + \boldsymbol{\Phi}_{rv}(t, t_0)\delta \boldsymbol{v}(t_0) \tag{3.47}$$

where $\boldsymbol{\Phi}_{rr}(t, t_0)$ and $\boldsymbol{\Phi}_{rv}(t, t_0)$ are corresponding block matrices of the state transition matrix. Substituting Eq. (3.47) into Eq. (3.46) yields

$$\begin{aligned} \phi_{\text{det}}(t) = \phi_0 + f_s(t - t_0) + \frac{f_s}{c}\boldsymbol{n} \cdot (\boldsymbol{\Phi}_{rr}(t, t_0) - \boldsymbol{I}_{3\times 3})\delta \boldsymbol{r}(t_0) \\ + \frac{f_s}{c}\boldsymbol{n} \cdot \boldsymbol{\Phi}_{rv}(t, t_0)\delta \boldsymbol{v}(t_0) + \frac{f_s}{c}\boldsymbol{n} \cdot (\tilde{\boldsymbol{r}}(t) - \tilde{\boldsymbol{r}}(t_0)). \end{aligned} \tag{3.48}$$

Given that $\boldsymbol{\Phi}_{rr}(t, t_0)$ and $\boldsymbol{\Phi}_{rv}(t, t_0)$ in Eq. (3.48) can be expressed as a Taylor expansion, Eq. (3.48) can be rewritten as

$$\begin{aligned} \phi_{\text{det}}(t) = \phi_0 + f_s(t - t_0) + \frac{f_s}{c}\sum_{k=1}^{\infty}\frac{(t - t_0)^k}{k!}\left[\boldsymbol{n}^T \boldsymbol{\varphi}_k \delta \boldsymbol{r}(t_0) + \boldsymbol{n}^T \boldsymbol{\gamma}_k \delta \boldsymbol{v}(t_0)\right] \\ + \frac{f_s}{c}\boldsymbol{n} \cdot (\tilde{\boldsymbol{r}}(t) - \tilde{\boldsymbol{r}}(t_0)). \end{aligned} \tag{3.49}$$

By defining $f_{orbit}^{(k)} = \frac{1}{(k)!}\frac{f_s}{c}\left[\boldsymbol{n}^T \boldsymbol{\varphi}_k \delta \boldsymbol{r}(t_0) + \boldsymbol{n}^T \boldsymbol{\gamma}_k \delta \boldsymbol{v}(t_0)\right]$, Eq. (3.49) can be transformed into

$$\phi_{\text{det}}(t) = \phi_0 + f_s(t - t_0) + \sum_{k=1}^{\infty} f_{orbit}^{(k)}(t - t_0)^k + \frac{f_s}{c}\boldsymbol{n} \cdot (\tilde{\boldsymbol{r}}(t) - \tilde{\boldsymbol{r}}(t_0)). \tag{3.50}$$

From Eq. (3.50), the improved phase propagation model can be divided into the predicted phase $\phi_{pre}(t) = f_s(t - t_0) + \frac{f_s}{c}\boldsymbol{n} \cdot (\tilde{\boldsymbol{r}}(t) - \tilde{\boldsymbol{r}}(t_0))$ and the phase to be estimated $\phi_{est}(t) = \phi_0 + \sum_{k=1}^{\infty} f_{orbit}^{(k)}(t - t_0)^k$.

For Earth-orbiting spacecraft, $\phi_{pre}(t)$ can be further transformed into

$$\phi_{pre}(t) = f_s(t - t_0) + \frac{f_s}{c}\boldsymbol{n} \cdot \left(\tilde{\boldsymbol{r}}_{SC/E}(t) - \tilde{\boldsymbol{r}}_{SC/E}(t_0)\right) + \frac{f_s}{c}\boldsymbol{n} \cdot (\tilde{\boldsymbol{r}}_E(t) - \tilde{\boldsymbol{r}}_E(t_0)) \quad (3.51)$$

where $\tilde{\boldsymbol{r}}_E(\cdot)$ is the position of the Earth which can be obtained by JPL DE series ephemeris. For estimation of ϕ_0 and $f_{orbit}^{(k)}(1 \le k \le \infty)$, the maximum K of k needs to be determined.

3.4.2.2 Index for Assessing Linearization Error

The polynomial $\sum_{k=1}^{K} f_{orbit}^{(k)}(t - t_0)^k$ is an expression used for approximating f_{oe} by means of linearization. f_{oe} is given as

$$f_{oe} = \frac{f_s}{c}\left[\boldsymbol{n}^T(\boldsymbol{\Phi}_{rr}(t, t_0) - \boldsymbol{I}_{3\times3})\delta\boldsymbol{r}(t_0) + \boldsymbol{n}^T\boldsymbol{\Phi}_{rv}(t, t_0)\delta\boldsymbol{v}(t_0)\right]. \quad (3.52)$$

Considering that exact solution of state transition matrix can be obtained from integration, then exact solution of f_{oe} is given by

$$\bar{f}_{oe} = \frac{f_s}{c}\left[\boldsymbol{n}^T\left(\overline{\boldsymbol{\Phi}}_{rr}(t, t_0) - \boldsymbol{I}_{3\times3}\right)\delta\boldsymbol{r}(t_0) + \boldsymbol{n}^T\overline{\boldsymbol{\Phi}}_{rv}(t, t_0)\delta\boldsymbol{v}(t_0)\right] \quad (3.53)$$

where $\overline{\boldsymbol{\Phi}}_{rr}(t, t_0)$ and $\overline{\boldsymbol{\Phi}}_{rv}(t, t_0)$ are exact solutions of $\boldsymbol{\Phi}_{rr}(t, t_0)$ and $\boldsymbol{\Phi}_{rv}(t, t_0)$.

The index defined below can be used to measure the truncation error and to determine K.

$$E_K = \left(1 - \frac{1}{\bar{f}_{oe}}\sum_{k=1}^{K} f_{orbit}^{(k)}(t - t_0)^k\right) \times 100\% \quad (3.54)$$

Sections 3.4.2.3 and 3.4.2.4 analyze E_K for different types of spacecraft.

3.4.2.3 Linearized Phase Propagation Model for Earth-Orbiting Spacecraft

Table 3.4 lists orbital elements of the investigated Earth-orbiting spacecraft. For $K = 1$ and $K = 2$, Figs. 3.18 and 3.19 show E_K with respect to semi-major axis of the spacecraft. It should be noted that data points in the aforesaid figures are the maximum E_K for different combinations of inclination and right ascension of ascending node with the same semi-major axis. The observation period of the pulsar is 3000 s.

Table 3.4 Initial orbital elements of near-Earth orbit spacecraft

Orbital element	Value
Semi-major axis [km]	[30000, 42000]
Eccentricity	0.0063
Inclination [°]	[0, 180)
Right ascension of ascending [°]	[0, 360)
Argument of perigee [°]	172.49
Mean anomaly [°]	43.73

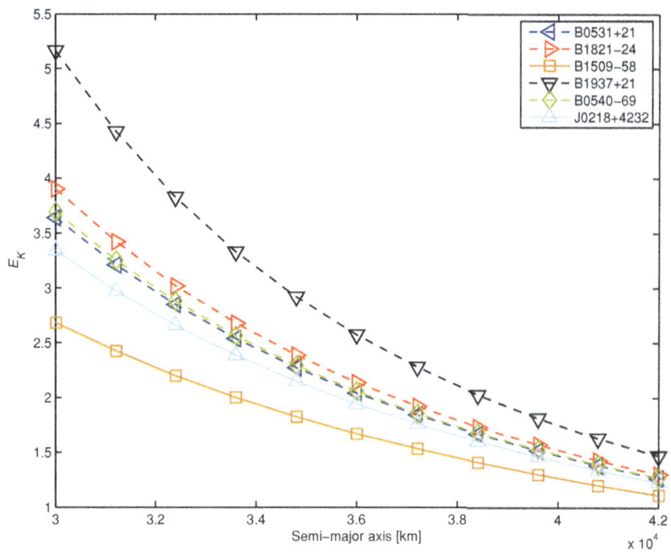

Fig. 3.18 Index versus semi-major axis with $K = 1$

For $K = 2$ with the orbital semi-major axis > 34000 km, E_K is higher than 2.5% for most of pulsars. Therefore, $K = 2$ can be applied for spacecraft orbiting on the high Earth orbits.

Then, Eq. (3.50) is transformed into

$$\begin{aligned}\phi_{\det}(t) &= \phi_0 + \left(f_s + f_{orbit}^{(1)}\right)(t - t_0) + f_{orbit}^{(2)}(t - t_0)^2 \\ &\quad + \frac{f_s}{c}\boldsymbol{n} \cdot \left(\tilde{\boldsymbol{r}}_{SC/E}(t) - \tilde{\boldsymbol{r}}_{SC/E}(t_0)\right) + \frac{f_s}{c}\boldsymbol{n} \cdot (\tilde{\boldsymbol{r}}_E(t) - \tilde{\boldsymbol{r}}_E(t_0)).\end{aligned} \tag{3.55}$$

Equation (3.55) represents the linearized phase propagation model for Earth-orbiting spacecraft.

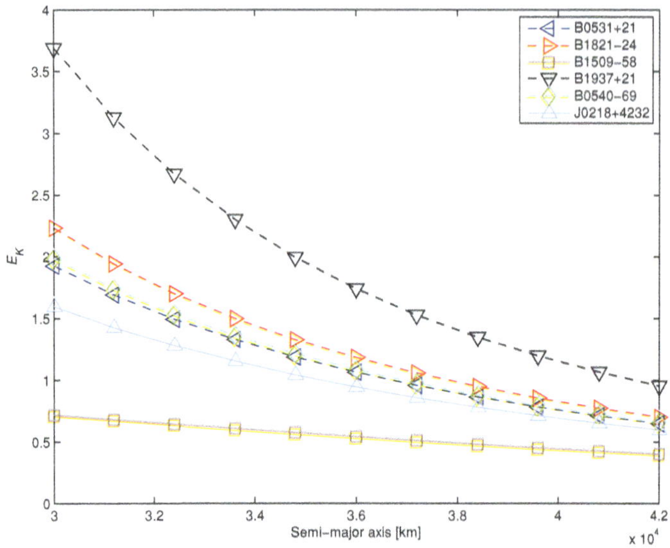

Fig. 3.19 Index versus semi-major axis with $K = 2$

3.4.2.4 Linearized Phase Propagation Model for Deep Space Explorers

Table 3.5 shows the orbital elements of deep space explorers.

For $K = 1$, Fig. 3.20 shows E_K relative to semi-major axis of the spacecraft. Note that data in Fig. 3.20 is the maximum E_K for different combinations of inclination and right ascension of ascending node with the same semi-major axis. The observation period of the pulsar is 7200 s.

As shown in Fig. 3.20, the index is on the order of 10^{-10} even for $K = 1$, which indicates that the first order is sufficient for the polynomial model. Therefore, the linearized phase propagation model for deep space explorers can be expressed as

$$\phi_{\det}(t) = \phi_0 + \left(f_s + f_{orbit}^{(1)}\right)(t - t_0) + \frac{f_s}{c}\boldsymbol{n} \cdot (\tilde{\boldsymbol{r}}(t) - \tilde{\boldsymbol{r}}(t_0)). \tag{3.56}$$

Table 3.5 Initial orbital elements of deep space explorers

Orbital elements	Value
Semi-major axis [km]	[1, 4]
Eccentricity	0.001
Inclination [°]	[0, 180)
Right ascension of ascending [°]	[0, 360)
Argument of perigee [°]	52.5
Mean anomaly [°]	60.34

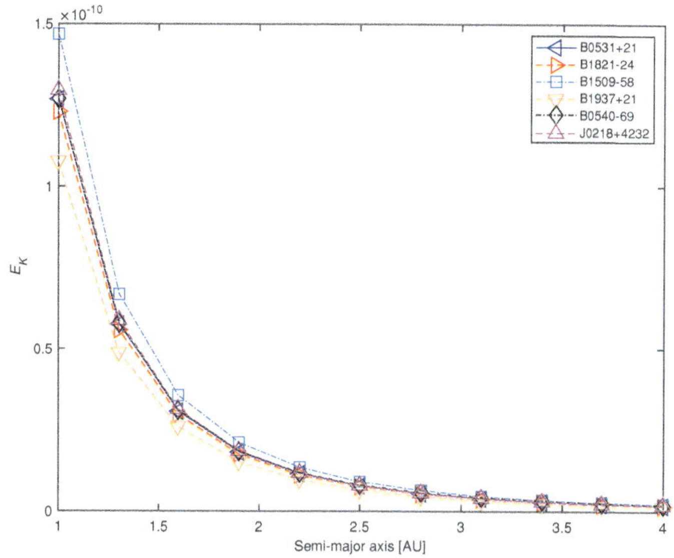

Fig. 3.20 E_K versus semi-major axis

3.4.3 Estimation of Phase and Doppler Frequency

A comparison between Eqs. (3.55), (3.56) and (3.42), (3.44) shows that approximation of nonlinear phase propagation model is achievable by simply estimating $\phi_0, f_{orbit}^{(1)}, f_{orbit}^{(2)}$ for linearized phase propagation model. Estimation of these values is equivalent to the estimation of the initial phase and Doppler frequency. There are two methods available for this purpose: direct use of photon TOAs and on-orbit epoch folding.

3.4.3.1 Direct Use of Photon TOAs

This method is developed based on direct use of photon TOAs. Substituting Eqs. (3.55) and (3.56) into Eq. (3.36) yields

$$LLF = \sum_{i=1}^{M} \ln\left[\begin{array}{l} \phi_0 + \left(f_s + f_{orbit}^{(1)}\right)(t_i - t_0) + f_{orbit}^{(2)}(t_i - t_0)^2 \\ + \frac{f_s}{c}\boldsymbol{n} \cdot \left(\tilde{\boldsymbol{r}}_{SC/E}(t_i) - \tilde{\boldsymbol{r}}_{SC/E}(t_0)\right) + \frac{f_s}{c}\boldsymbol{n} \cdot \left(\tilde{\boldsymbol{r}}_E(t_i) - \tilde{\boldsymbol{r}}_E(t_0)\right) \end{array}\right] \tag{3.57}$$

and

$$LLF = \sum_{i=1}^{M} \ln\left[\phi_0 + \left(f_s + f_{orbit}^{(1)}\right)(t_i - t_0) + \frac{f_s}{c}\boldsymbol{n}\cdot(\tilde{\boldsymbol{r}}(t_i) - \tilde{\boldsymbol{r}}(t_0))\right]. \tag{3.58}$$

The estimation of $\phi_0, f_{orbit}^{(1)}$ and $f_{orbit}^{(2)}$ can be obtained by searching for the maximum of Eqs. (3.57) and (3.58).

Considering that the computational complexity of this method is proportional to the number of photons received over the observation period, it is recommended to be used for data of millisecond pulsars.

For Eq. (3.55), CRB for estimation of phase and Doppler frequency can be derived. The following pulsar signal rate model is considered based on Eq. (3.55)

$$\lambda(t) = \alpha h\left(\begin{array}{l}\phi_0 + f_{sdy}(t - t_0) + f_{orbit}^{(2)}(t - t_0)^2 \\ + \frac{f_s}{c}\boldsymbol{n}\cdot\left(\tilde{\boldsymbol{r}}_{SC/E}(t) - \tilde{\boldsymbol{r}}_{SC/E}(t_0)\right) + \frac{f_s}{c}\boldsymbol{n}\cdot(\tilde{\boldsymbol{r}}_E(t) - \tilde{\boldsymbol{r}}_E(t_0))\end{array}\right) + \beta \tag{3.59}$$

where $f_{sdy} = f_s + f_{orbit}^{(1)}$.

Then, CRB of $\phi_0, f_{sdy}, f_{orbit}^{(2)}$ can be obtained from the following lemma.

Lemma 1 *Assuming*

$$\boldsymbol{\theta} \triangleq \left[\phi_0\; f_{sdy}\; f_{orbit}^{(2)}\right]^T \tag{3.60}$$

Then CRB of $\boldsymbol{\theta}$ *in* Eq. (3.60) *is given by*

$$CRB(\boldsymbol{\theta}) = \boldsymbol{IM}^{-1/2}(\boldsymbol{\theta}) \tag{3.61}$$

where

$$\boldsymbol{IM}(\boldsymbol{\theta}) = \begin{bmatrix} I_{\phi_o} & I_{\phi_o f_{sdy}} & I_{\phi_o f_{orbit}^{(2)}} \\ I_{\phi_o f_{sdy}} & I_{f_{sdy}} & I_{f_{sdy} f_{orbit}^{(2)}} \\ I_{\phi_o f_{orbit}^{(2)}} & I_{f_{sdy} f_{orbit}^{(2)}} & I_{f_{orbit}^{(2)}} \end{bmatrix} \tag{3.62}$$

where

$$I_{\phi_o} = \int_0^{T_{obs}} \frac{\left[\alpha h'(\phi_{\det}(t))\right]^2}{h(\phi_{\det}(t))}dt, \quad I_{f_{sdy}} = \int_0^{T_{obs}} \frac{t^2\left[\alpha h'(\phi_{\det}(t))\right]^2}{h(\phi_{\det}(t))}dt,$$

$$I_{f_{orbit}^{(2)}} = \int_0^{T_{obs}} \frac{t^4\left[\alpha h'(\phi_{\det}(t))\right]^2}{h(\phi_{\det}(t))}dt, \quad I_{\phi_o f_{sdy}} = \int_0^{T_{obs}} \frac{t\left[\alpha h'(\phi_{\det}(t))\right]^2}{h(\phi_{\det}(t))}dt,$$

$$I_{\phi_0 f_{orbit}^{(2)}} = \int_0^{T_{obs}} \frac{t^2 \left[\alpha h'(\phi_{\det}(t))\right]^2}{h(\phi_{\det}(t))} dt, \quad I_{f_{sdy} \cdot f_{orbit}^{(2)}} = \int_0^{T_{obs}} \frac{t^3 \left[\alpha h'(\phi_{\det}(t))\right]^2}{h(\phi_{\det}(t))} dt.$$

Since Eq. (3.61) is just an expansion of theorem 4.2 in Ref. [3], the proof can be omitted.

3.4.3.2 On-Orbit Epoch Folding

Besides direct use of photon TOAs, on-orbit epoch folding is presented for Eq. (3.56), which proceeds by three steps: (1) removal of spacecraft orbital motion effect; (2) search for the period of converted photon TOAs; and (3) estimation of initial phase.

(1) **Removal of spacecraft orbital motion effect**

When $K = 1$, $\boldsymbol{\Phi}(t, t_0)$ can be given as

$$\boldsymbol{\Phi}(t, t_0) = \boldsymbol{I}_{6\times6} + \boldsymbol{G}(t - t_0) \tag{3.63}$$

where

$$\boldsymbol{G} = \begin{bmatrix} \boldsymbol{0}_{3\times3} & \boldsymbol{I}_{3\times3} \\ \boldsymbol{S} & \boldsymbol{0}_{3\times3} \end{bmatrix}. \tag{3.64}$$

In Eq. (3.64), $\boldsymbol{S}$ is expressed as

$$\boldsymbol{S} = \left.\frac{\partial \dot{\boldsymbol{v}}}{\partial \boldsymbol{r}}\right|_t. \tag{3.65}$$

Equation (3.52) is transformed into

$$f_{oe} = f_{orbit}^{(1)} = \frac{f_s}{c} \boldsymbol{n} \cdot \delta \boldsymbol{v}(t_0). \tag{3.66}$$

From Eq. (3.56),

$$t - t_0 + \frac{1}{c} \frac{f_s}{(f_s + f_{orbit}^{(1)})} \boldsymbol{n} \cdot (\tilde{\boldsymbol{r}}(t) - \tilde{\boldsymbol{r}}(t_0)) = \frac{1}{f_s + f_{orbit}^{(1)}} \phi_{\det}(t) - \frac{1}{f_s + f_{orbit}^{(1)}} \phi_0. \tag{3.67}$$

Substituting Eq. (3.66) into Eq. (3.67) yields

$$\begin{aligned} & t - t_0 + \frac{1}{c} \frac{f_s}{(f_s + f_{orbit}^{(1)})} \boldsymbol{n} \cdot (\tilde{\boldsymbol{r}}(t) - \tilde{\boldsymbol{r}}(t_0)) \\ & = t - t_0 + \frac{1}{c} \frac{c}{(c + \boldsymbol{n} \cdot \delta \boldsymbol{v}(t_0))} \boldsymbol{n} \cdot (\tilde{\boldsymbol{r}}(t) - \tilde{\boldsymbol{r}}(t_0)) \end{aligned}$$

$$\approx t - t_0 + \frac{1}{c}\boldsymbol{n} \cdot (\tilde{\boldsymbol{r}}(t) - \tilde{\boldsymbol{r}}(t_0)). \tag{3.68}$$

By defining $\bar{t} = \phi_{\text{det}}(t)\left(f_s + f_{orbit}^{(1)}\right)^{-1}$ and $\bar{t}_0 = \phi_0\left(f_s + f_{orbit}^{(1)}\right)^{-1}$, Eq. (3.68) can be transformed into

$$t - t_0 + \frac{1}{c}\boldsymbol{n} \cdot (\tilde{\boldsymbol{r}}(t) - \tilde{\boldsymbol{r}}(t_0)) = \bar{t} - \bar{t}_0. \tag{3.69}$$

Appling Eq. (3.69) to the photon TOAs could convert $\{t_i\}_{i=1}^{M}$ with a time-variant frequency into $\{\bar{t}_i\}_{i=1}^{M}$ with unknown constant frequency. In other words, Eq. (3.69) can remove the spacecraft orbital effect on the original photon pulse TOAs. The period of $\{\bar{t}_i\}_{i=1}^{M}$ is $\overline{P} = \left(f_s + f_{orbit}^{(1)}\right)^{-1}$.

(2) **Search for the period of converted photon TOAs $\{\bar{t}_i\}_{i=1}^{M}$**

In this step, any period search method can be used for searching for $\overline{P}$, and the likelihood function method given in Sect. 3.2.2 is adopted herein.

(3) **Estimation of initial phase**

When the period $\widehat{\overline{P}}$ is found out, a new empirical profile can be recovered by using $\widehat{\overline{P}}$. Therefore, the initial phase can be obtained by the current phase estimation method provided in Sect. 3.3.1.1. The fast near-maximum likelihood method shown in Eq. (3.30) is adopted in this paper, because it is computationally efficient.

(4) **Computation complexity of on-orbit epoch folding**

The on-orbit epoch folding method includes three steps. Step 1 requires M additions and multiplications. Assuming there are M_p potential periods, it requires $M_p N_{fold} M$ additions and multiplications to fulfill Step 2. Based on Ref. [3], the fast near-maximum likelihood method requires $N_b(1 + 2\log_2 N_b)$ additions and multiplications. Thus, the whole computation complexity of the proposed on-orbit epoch folding is approximated as $O(M_p N_{fold} M + M + N_b(1 + 2\log_2 N_b))$.

From Eq. (3.69), the period of $\{\bar{t}_i\}_{i=1}^{M}$ is constant. Therefore, when the first step is done, the initial phase and frequency of $\{\bar{t}_i\}_{i=1}^{M}$ can also be estimated by the mean of direct use of photon TOAs. Then, the whole computation complexity becomes $O(6MN_f N_\phi + M)$, where, N_ϕ and N_f are grid numbers for phase estimation and frequency estimation, respectively. When $N_\phi = N_b$ and $N_f = M_p$, the computational complexity of the proposed method is much less than that of the case where the direct use of photon TOAs is adopted. Thus, the proposed on-orbit epoch folding is more practical.

3.4.4 *Simulation Analysis*

To verify the performance of the dynamic pulsar signal processing method based on orbital dynamics, two cases are considered in this section: (1) observation of millisecond pulsars for near-earth orbit spacecraft, and (2) observation of young pulsars for deep space explorers.

3.4.4.1 Simulation for Earth-Orbiting Spacecraft Observing Millisecond Pulsars

In this case, the initial phase and Doppler frequency of pulsar signals are calculated via the direct use of the photon TOAs. The millisecond pulsar PSR B1821−24 is adopted for simulation, with simulation parameters listed in Table 3.2. Table 3.6 shows the initial orbit elements of high-Earth orbit spacecraft, and the initial orbital errors are 100 m and 0.1 m/s.

Figure 3.21 illustrates the root mean square (RMS) of initial phase estimation varying with the observation period. For observation period <100 s, RMS is a non-confidence interval; for observation period >100 s, RMS is consistent with CRB; and for observation period of 3000 s, RMS finally reaches 0.001.

Figure 3.22 shows the pulsar TOA relative to observation period. Similar with the phase estimation, the pulse TOA converges to CRB after the observation period is longer than 100 s, and afterwards the method performance improves over observation period.

We will further analyze the factors that might affect the performance of proposed method when the observation period is assumed to be 1000 s.

Figures 3.23 and 3.24 show how the initial position error and velocity error affect the RMS of initial phase estimation. Although the increase of position and velocity errors will degrade the method performance, the magnitude of the increase of RMS is less than 1×10^{-4} even when position and velocity errors are 1000 m and 5 m/s, respectively. Therefore, this method can be considered insensitive to the initial state error of spacecraft.

Table 3.6 Initial parameters of high-Earth orbit spacecraft

Orbital elements	
Semi-major axis [km]	34000
Eccentricity	0.0063
Inclination [°]	63.33
Right ascension of ascending node [°]	263.55
Argument of perigee [°]	172.49
Mean anomaly [°]	43.73

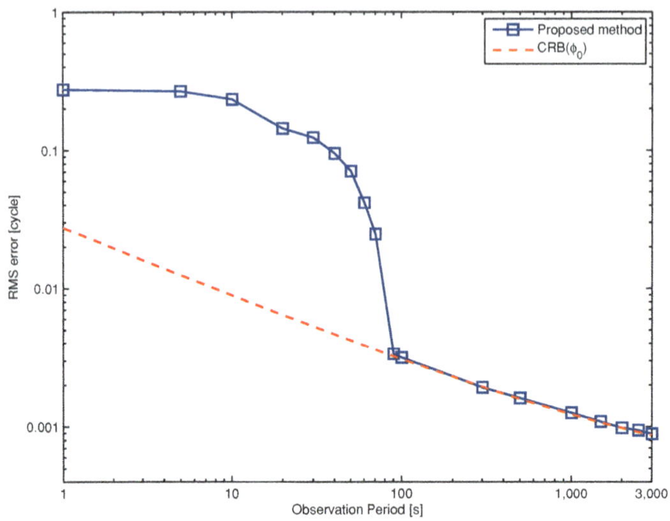

Fig. 3.21 Initial phase estimation error versus observation period

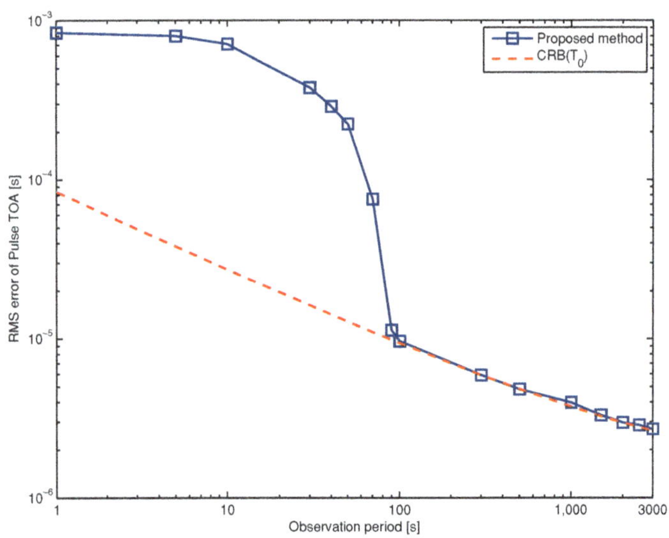

Fig. 3.22 Pulse TOA estimation error versus observation period

Figure 3.25 shows the variation of RMS with respect to semi-major axis, in which RMS slightly decreases with the semi-major axis. This means the method performance is slightly affected by the change of semi-major axis.

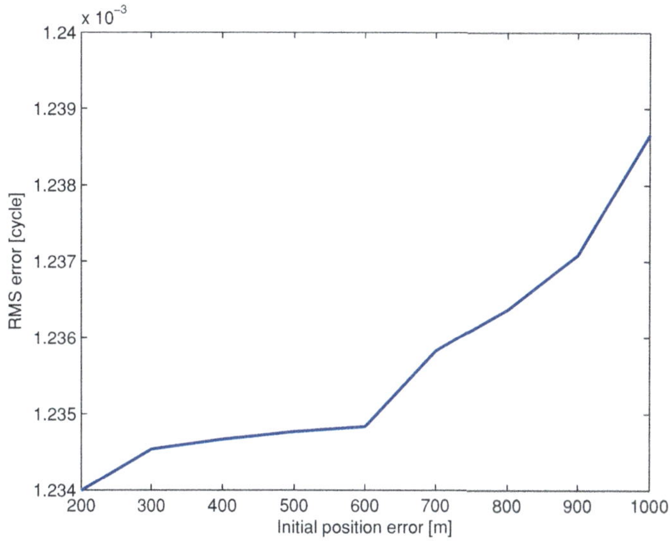

Fig. 3.23 Initial phase estimation error versus initial position error of spacecraft

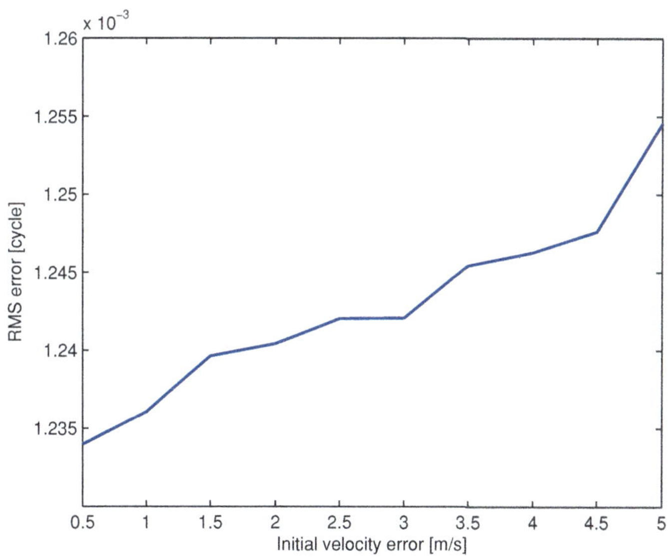

Fig. 3.24 Initial phase estimation error versus initial velocity error of spacecraft

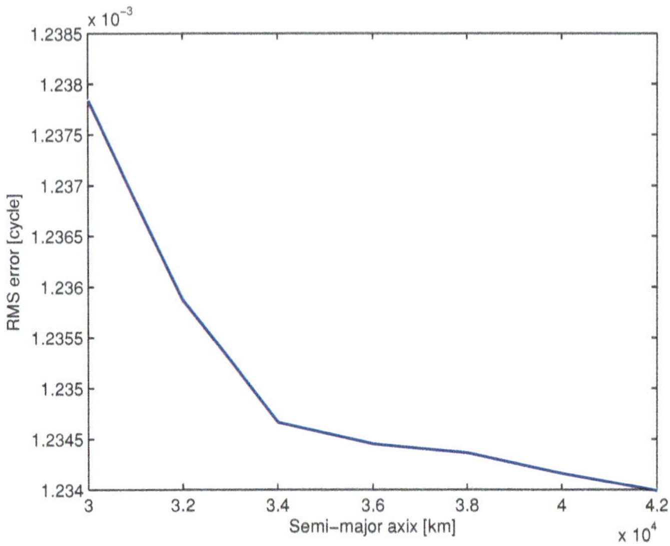

Fig. 3.25 Initial phase estimation error versus semi-major axis of spacecraft

The computation complexity of the algorithm is measured by the computing time of CPU, with simulation environment of CPU: Intel I7-4710MQ @ 2.5 GHz, compiling environment of Visual studio 6.0, and 10 processes of parallel computing. Figure 3.26 shows the CPU time varying with the observation period. CPU time

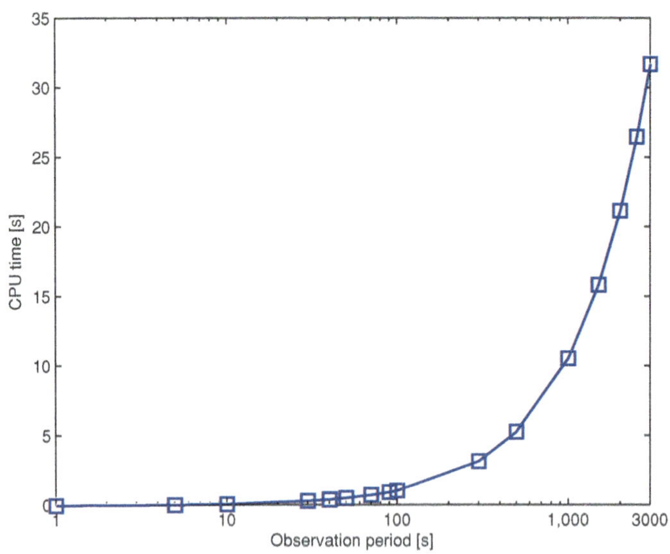

Fig. 3.26 CPU time versus observation period

increases over the observation period, because that the computation complexity of the direct use of photon TOAs method is in direct proportion to the photon number which increases over the observation period. For observation period of 3000 s, CPU time is only 35 s which can be neglected when compared to the observation period.

3.4.4.2 Simulation for Deep Space Explorers Observing Young Pulsars

In this case, the initial phase and Doppler frequency are calculated by on-orbit epoch folding method. PSR B0531+21 is for investigation, with profile as illustrated in Fig. 3.2 and simulation parameters listed in Table 3.1. Table 3.7 gives the orbital elements of the spacecraft with initial position error and velocity error of 100 km and 10 m/s. For accurate estimation of initial phase, N_{fold} and N_b are assumed to be 2000 and 33400, respectively.

With the spacecraft motion effect is removed, Fig. 3.27 provides the estimation of $\hat{\phi}_0$ for two different cases: (1) $f_{orbit}^{(1)}$ is estimated, and (2) $f_{orbit}^{(1)}$ is ignored. In the latter case, the profile is folded by directly using the pulsar rotation period.

For observation period <100 s, both cases have similar results which approach the CRB. This is because the phase accumulated as result of presence of $f_{orbit}^{(1)}$ is less than the width of a bin, namely, 2.99×10^{-5}. However, these two RMS curves start to separate from each other when observation period >100 s. The curve in Case 1 becomes divergent while the one in Case 2 keeps on converging, which is due to the fact that $f_{orbit}^{(1)}$ gives a gradually significant effect as the observation period increases. Consequently, $f_{orbit}^{(1)}$ is negligible for observation period <100 s, and must be taken into account if the period increases. An observation period of 100 s provides a CRB of 3.37×10^{-4}, a pulse TOA accuracy of 1.26 μs and a position error of 3.38 km. So the observation period needs to be extended for high-precision navigation result, which also manifests the need for estimating $f_{orbit}^{(1)}$.

Figure 3.28 illustrates the estimation error of $\hat{f}_{orbit}^{(1)}$ in Case 2, which increases over the observation period. This suggests that the performance of the period search method depends on the number of photons. For observation period of 3000 s, RMS equals to $3 \times 10^{-7}\text{s}^{-1}$. From Eq. (3.66), the velocity estimation accuracy in such case is about 1 m/s. Although it seems undesirable, $\hat{f}_{orbit}^{(1)}$ is just for refining the solution

Table 3.7 Initial orbital elements of deep space explorers

Orbital elements	Value
Semi-major axis [km]	1.94
Eccentricity	0.01
Inclination [°]	23.56
Right ascension of ascending node [°]	233.55
Argument of perigee [°]	132.49
Mean anomaly [°]	33.73

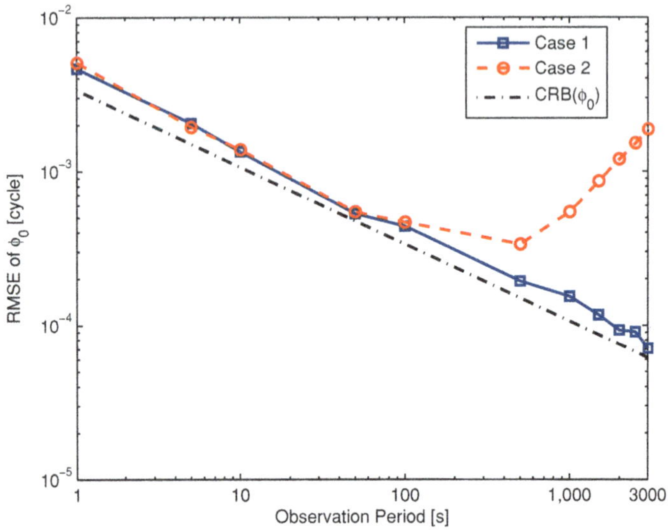

Fig. 3.27 Initial phase estimation error changes in two cases

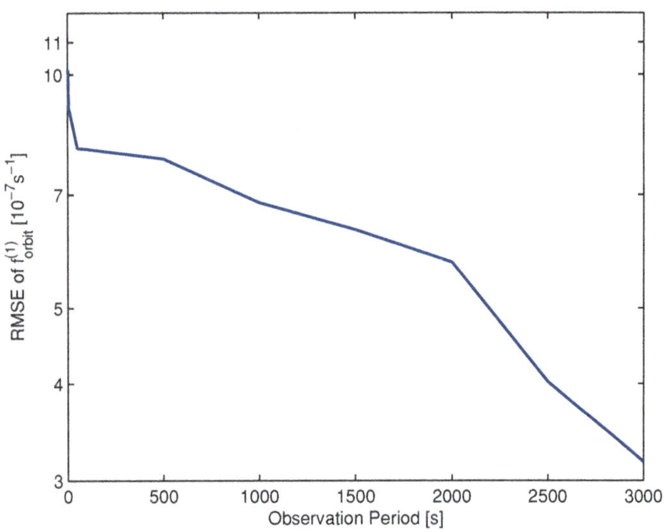

Fig. 3.28 $f_{orbit}^{(1)}$ estimation error

of initial phase and pulse TOA. The navigation filter is recommended for estimating the spacecraft velocity, as it can provide a better estimation result by considering the orbital dynamics of spacecraft. In addition, when the initial velocity error is less than 1 m/s, its influence over the initial phase estimation is negligible. In such case, the profile may be recovered by using the pulsar spinning period.

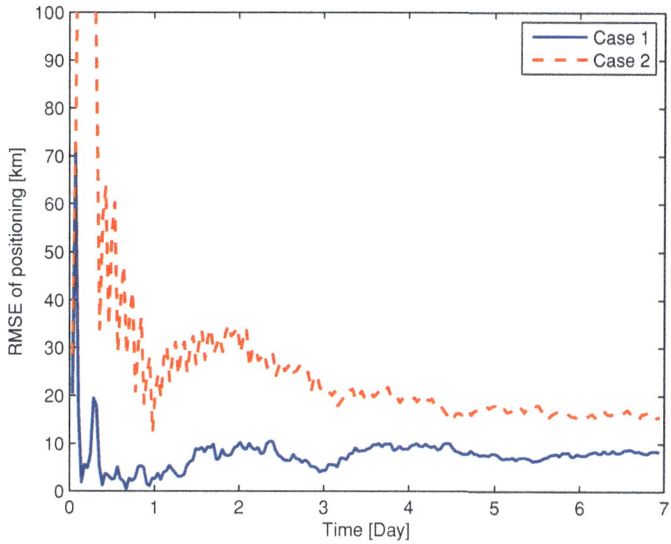

Fig. 3.29 Comparison of position estimation errors in both cases

The navigation positioning performance is further analyzed for both cases. Besides PSR B0531+21, PSR B1821−24 and PSR B1937+21 are also employed in navigation simulation with observation period assumed to be 3000 s. A gimbal is assumed on the deep space explorer for sequential observation of three pulsars.

Figures 3.29 and 3.30 show the estimation error of position and velocity in these two cases. Although the position and velocity estimation errors converge as the observation period increases in both cases, Case 1 provides higher rate and precision of convergence, so $f_{orbit}^{(1)}$ needs to be considered.

The computation complexity of the algorithm is measured by the computing time of CPU, with simulation environment of CPU: Intel I7-4710MQ @ 2.5 GHz, compiling environment of Visual studio 6.0, and 10 processes of parallel computing adopted in period search procedures.

Figure 3.31 shows the computing time of CPU. As we can see, the CPU time increases over observation period, which is mainly consumed for period searching, because TOAs of all photons need to be repeatedly used for many times. For observation period of 3000 s, CPU time equals to 25 s. Considering the CPU time is much less than the observation period (generally several hours) of deep space explorer, the computation complexity of the algorithm is acceptable for the deep space explorer.

Figures 3.32 and 3.33 analyze the effect of initial velocity error and semi-major axis of the spacecraft on $\hat{\phi}_0$ for observation period of 3000 s. When the initial velocity error increases from 20 m/s to 100 m/s, $\hat{\phi}_0$ estimation error increases slowly. Deep space explorers typically have initial velocity error of no greater than 50 m/s. Therefore, the on-orbit epoch folding method is applicable to autonomous navigation of

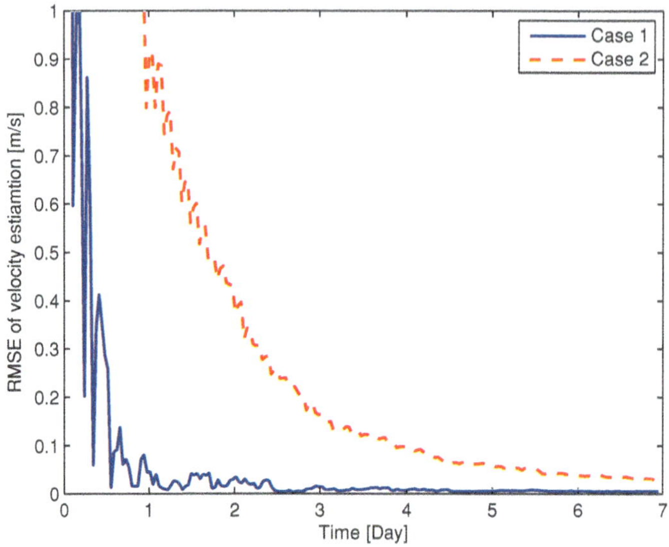

Fig. 3.30 Comparison of velocity estimation errors in both cases

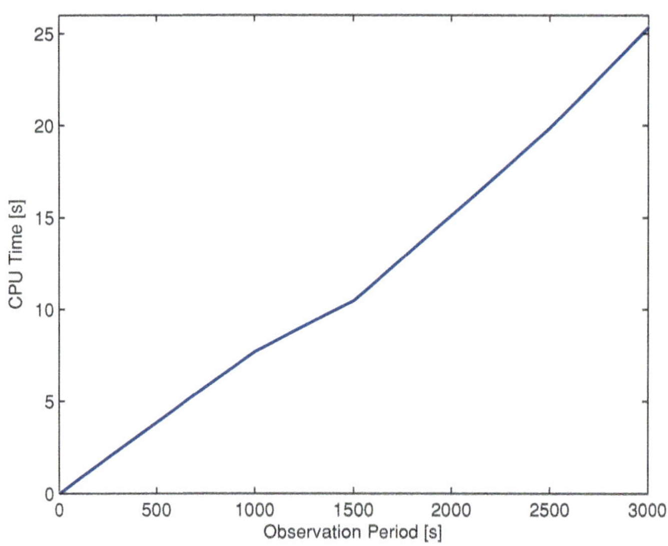

Fig. 3.31 CPU time for on-orbit epoch folding

deep space explorers. $\hat{\phi}_0$ estimation error slightly fluctuates with the increase of semi-major axis. So the on-orbit epoch folding method is insensitive to the semi-major axis and applicable to various applications.

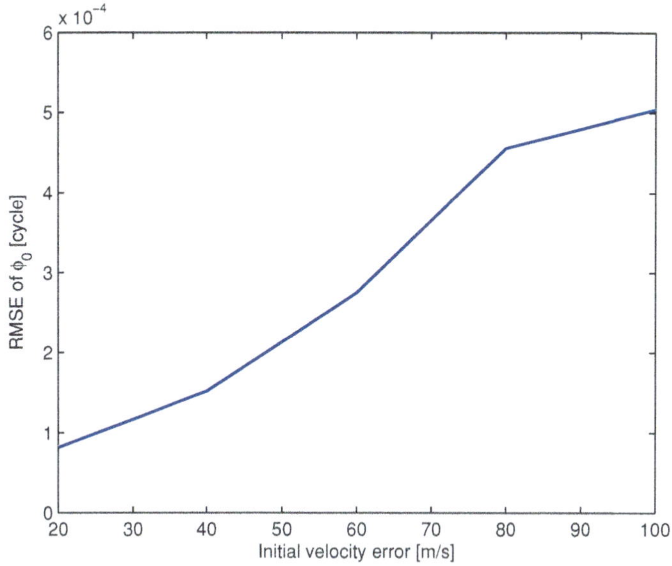

Fig. 3.32 Influence of initial velocity error of spacecraft

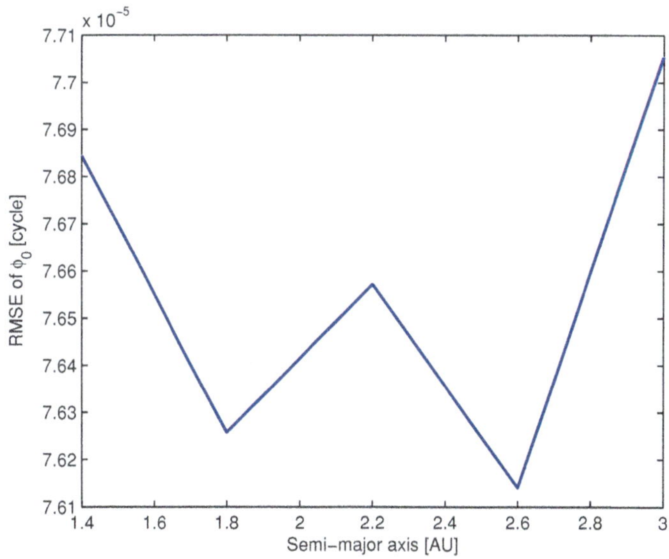

Fig. 3.33 Influence of orbital semi-major axis

3.5 Data Processing of XPNAV-1 Data

3.5.1 Introduction of the Measured Data of XPNAV-1

3.5.1.1 Introduction of XPNAV-1

XPNAV-1 was launched by China on November 10th, 2016. There are several experiment tasks for XPNAV-1. Firstly, to verify performances of X-ray detector in the space environment. Secondly, to observe the pulsar PSR B0531+21 and recover its profile. The weight of this satellite is only over 200 kg, and two types of X-ray detectors are loaded, including the collimated MCP and the grazing incidence focusing and SDD detector.

3.5.1.2 Description of Data from XPNAV-01

On May 9th, 2017, Beidou satellite navigation system website with address of http://www.beidou.gov.cn/xpnavdata.rar released the first set of observation data of PSR B0531+21 from the grazing incidence focusing detector.

There are two types of files for the released data, i.e., the file for photon arrival events and the file for the satellite orbit. The detailed file formats are provided in Table 3.8. Meanwhile, parameters of the grazing incidence focusing detector are shown in Table 3.9.

Table 3.8 File format of the released data

	Photon arrival event file	Orbit file
File name format	PSRNAME_YYYYMMDD_HHMMSS.txt	ORBIT_YYYYMMDD_HHMMSS.txt
File naming rules	PSRNAME: Pulsar Name	ORBIT: Orbit File
	YYYY: Year of the Start Time in UTC	YYYY: Year of the Start Orbit in UTC
	MM: Month of the Start Time in UTC	MM: Month of the Start Orbit in UTC
	DD: Day of the Start Time in UTC	DD: Day of the Start Orbit in UTC
	HH: Hour of the Start Time in UTC	HH: Hour of the Start Orbit in UTC
	MM: Minute of the Start Time in UTC	MM: Minute of the Start Orbit in UTC
	SS: Second of the Start Time in UTC	SS: Second of the Start Orbit in UTC
File content description	Photon TOAs (Seconds relative to 00:00:00 January 1, 2008 in UTC)	Orbital Epoch (Seconds relative to 00:00:00 January 1, 2008 in UTC)
		J2000-X(m)
		J2000-Y(m)
		J2000-Z(m)
		J2000-VX(m/s)
		J2000-VY(m/s)
		J2000-VZ(m/s)

Table 3.9 Performance parameters of the grazing incidence focusing detector

Parameter	Indicator
Range of detection spectrum	0.5–10 keV
Optical filed of view (FOV)	$2\omega = 15'$
Geometric area	30cm2@1.5 keV
Time resolution	$\leq$1.5 μs
Energy resolution	$\leq$180 eV@5.9 keV

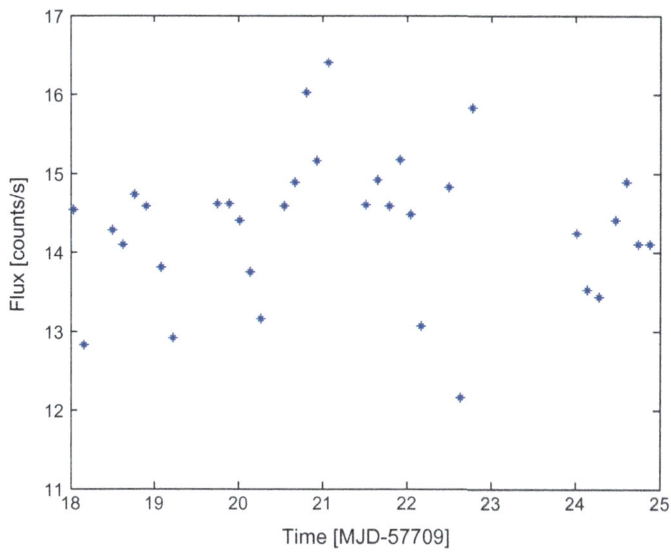

Fig. 3.34 Photon flow statistics for PSR B0531+21

3.5.2 Data Processing for the Measured Data

3.5.2.1 Photon Flow Statistics for PSR B0531+21

Analyze the data of PSR B0531+21 from November 17th to December 19th, 2016 (UTC). The total observation duration is 366720 s, and 4438661 photons are recorded (including detector background noise). Figure 3.34 shows the photon counting rate per minute of PSR B0531+21.

3.5.2.2 Distribution of Photon Time Interval for PSR B0531+21

Figure 3.35 shows the distribution of photon time interval for PSR B0531+21 using the data. The solid line in this figure indicates the distribution of measured data and

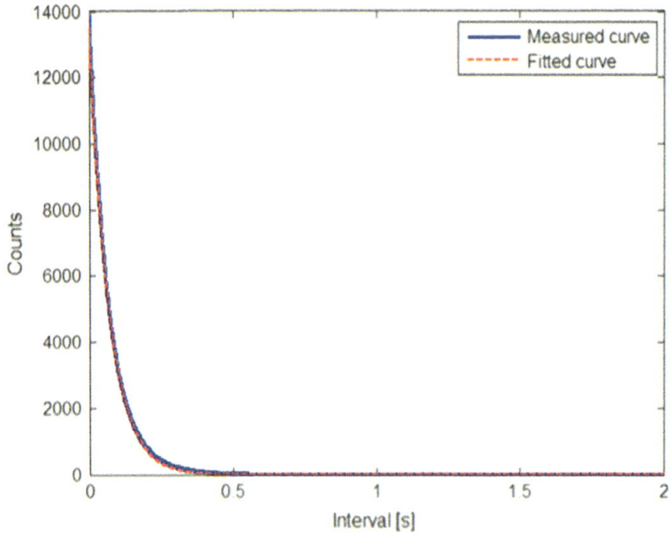

Fig. 3.35 Distribution of photon time interval for PSR B0531+21

the dashed one is the fitted curve. As the figure shown, photon time interval follows a negative exponential distribution, whose expression is given as

$$y = 13528.25 \cdot e^{-14.32x} \tag{3.70}$$

3.5.2.3 Energy Analysis

Using the data, the energy spectrum of 1–7.2 keV is established, and the curve are shown in Fig. 3.36.

3.5.2.4 Period Analysis for PSR B0531+21

Jodrell Bank Observatory published the period of PSR B0531+21, which is 0.033730087359680 s, on December 16th, 2016, while the period searched by using the data of XPNAV-1 on December 16th is 0.033730085994417 s. The difference between these two periods is 1.365263 ns.

Figure 3.37 shows the difference between frequencies of the measured data and Jodrell Bank Observatory from November 18th to December 19th, 2016. The mean of the period error is 1.992857 ns, and the standard variance is 15.130303 ns.

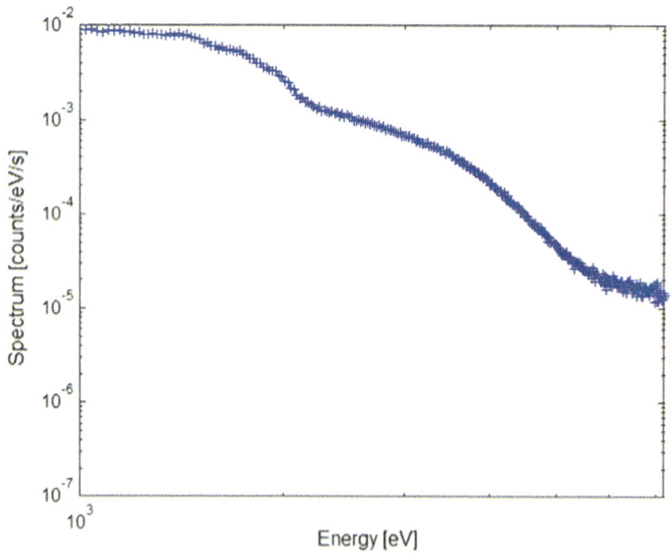

Fig. 3.36 Energy spectrum of 1–7.2 keV

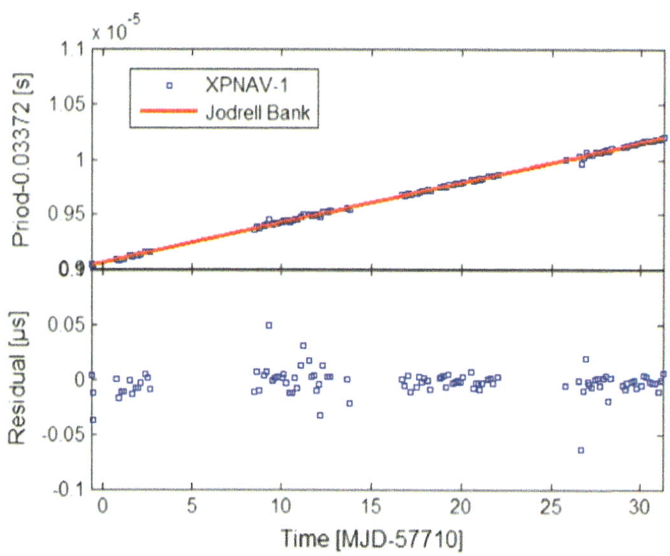

Fig. 3.37 Frequency comparison of the measured data and the data of Jodrell Bank Observatory

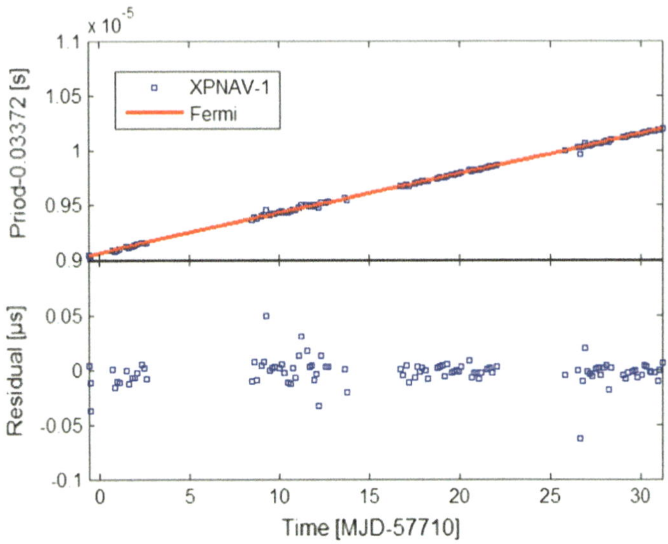

Fig. 3.38 Period comparison between the measured data of XPNAV-1 and Fermi

Figure 3.37 shows the difference between periods searched by the data of XPNAV-1 and Fermi from November 18th to December 19th, 2016. The mean of the period error is 0.974007 ns, and the standard variance is 16.323244 ns (Fig. 3.38).

3.5.2.5 Profile Recovery for PSR B0531+21

Using the data of PSR B0531+21 from November 18th to December 19th, 2016, a template can be established. Transferring these recorded photon TOAs from XPNAV-1 to SSB, the profile is recovered and shown in Figs. 3.39 and 3.40. They show the profile obtained by the data of XPNAV-01 is consistent with that obtained by RXTE.

3.5.2.6 Timing Analysis for PSR B0531+21

Using the data of PSR B0531+21 from November 18th to December 19th, 2016 and applying the pulsar-timing package tempo2, the timing result is obtained and shown in Fig. 3.41. As shown in this figure, the timing result of XPNAV-1 is consistent with that of Fermi, and its accuracy is higher than 200 μs.

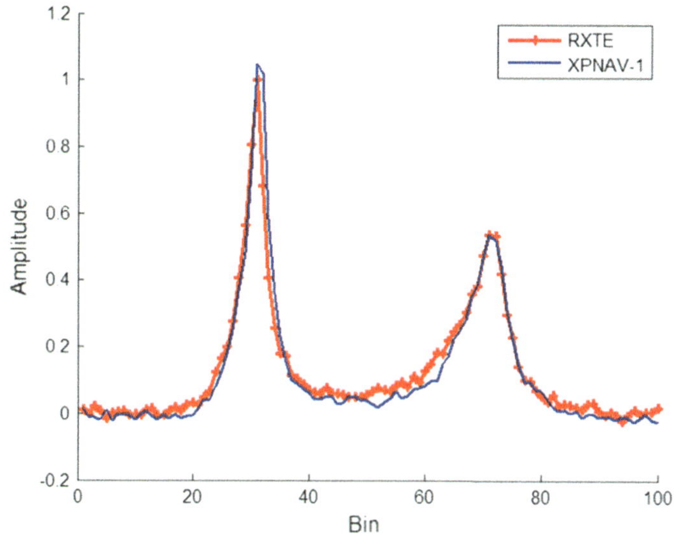

Fig. 3.39 Profile of PSR B0531+21 folded by the data from November 18th to December 19th, 2016 (100 bin)

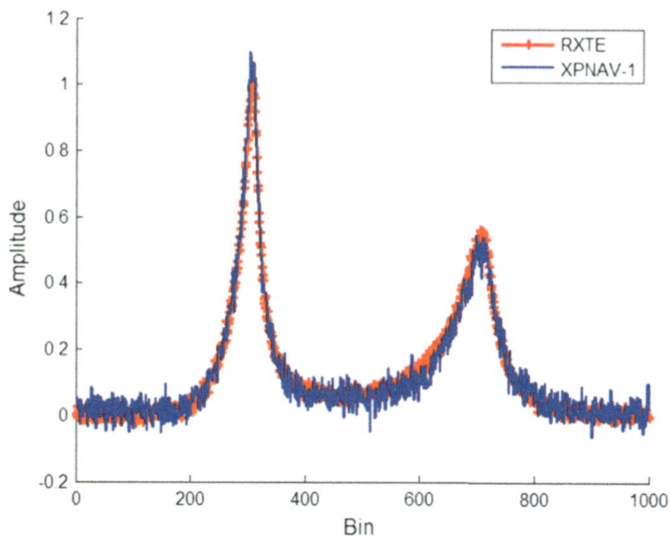

Fig. 3.40 Profile of PSR B0531+21 folded by the data form November 18th to December 19th, 2016 (1000 bin)

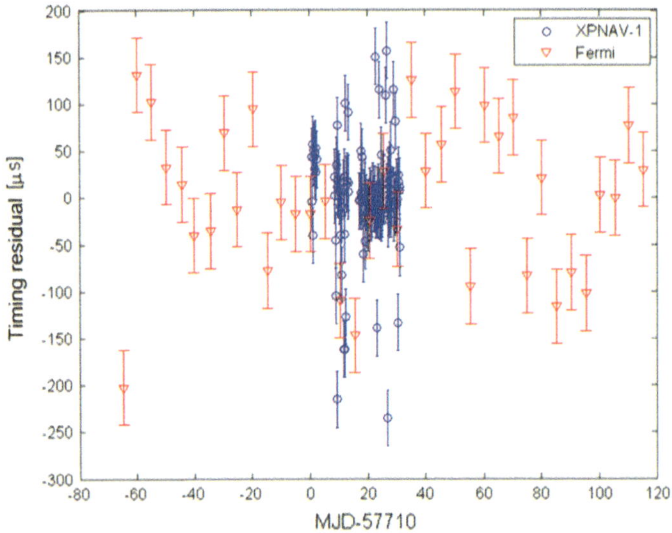

Fig. 3.41 Comparison between the timing results of XPNAV-1 and Fermi

3.6 Summary

To establish high-precision pulsar signal profile and process dynamic pulsar signals, this chapter studies on key techniques for building pulsar profile, and presents the dynamic pulsar signal processing methods based on orbital dynamics. Main work and conclusions are as follows:

(1) A generalized time-domain pulsar signal period search method is developed. A generalized period search framework is established by treating the pulsar signal period search as a convex optimization problem. Based on this framework, the indicator functions are provided for minimum entropy method and χ^2 search method, and the recovered profile is regarded as multi-dimensional independent distribution to derive and establish the period search method based on likelihood function. The influencing factors of pulsar signal period search are analyzed by means of simulation data, which indicates that increase of the photon number will improve the accuracy of period search. The period search method based on likelihood function is verified for effectiveness by means of RXTE satellite data.

(2) To improve SNR of recovered profile, the profile recovery method based on the optimum N_b and profile de-noising method based on kernel regression are presented. On one hand, the recovered profile is considered as an implicit function of N_b to establish the corresponding optimization problem. The indicator function for optimization problem is derived based on MISE by using the cross-validation method, of which the effectiveness is verified by means of RXTE satellite data. On the other hand, the recovered profile of pulsar is fitted based

on kernel regression with an appropriate bandwidth automatically selected via cross validation. The data simulation of two extreme pulsars and comparison with wavelet de-noising and EMD indicate that profile recovery based on kernel regression introduces the minimum phase deviation while enhancing SNR of profile. The data analysis of RXTE satellite suggests that kernel regression can retain more details of the recovered profile.

(3) To remove the impact of orbital motion of spacecraft on the pulsar signal processing, the pulsar signal processing method based on orbital dynamics is presented. Firstly, the phase propagation model for orbital motion of spacecraft is established by introducing the orbital dynamics model into the pulsar signal phase propagation model. Then, the phase propagation model is decomposed into the estimated phase and the phase to be estimated based on the fact that orbital dynamic model can provide an estimated position for orbital motions of spacecraft. The estimated phase may be extrapolated from orbital dynamics, while the phase to be estimated needs to be calculated. The three-order phase estimation model and its CRB are derived for near-Earth orbit spacecraft and low-flux pulsars. For deep space explorers and high-flux pulsars, the phase and Doppler frequency shift estimation framework is established to process dynamic signals by removing the impact of orbital motions, searching period and estimating initial phase. The effectiveness of this method is verified by simulation analysis.

References

1. Konishi S, Kitagawa G (2008) Information criteria and statistical modeling. Springer Press, Berlin
2. Graham M, Drake A, Djorgovski S et al (2013) A comparison of period finding algorithms. Mon Not R Astron Soc 434:3423–3444
3. Emadzadeh A, Speyer J (2011) Navigation in space by X-ray pulsars. Springer Press, Berlin
4. Takeda H, Farsiu S, Milanfar P (2007) Kernel regression for image processing and reconstruction. IEEE Trans Image Process 16:349
5. Arlot S (2010) A survey of cross-validation procedures for model selection. Stat Surv 4:40–79
6. Rinauro S, Colonnese S, Scarano G (2013) Fast near-maximum likelihood phase estimation of X-ray pulsars. Singal Process 93:326–331

Chapter 4
Errors Within the Time Transfer Model and Compensation Methods for Earth-Orbing Spacecraft

To essentially enhance the performance of the X-ray pulsar-based navigation system, this chapter systematically presents the error sources within the time transfer model and analyzes their impacts on the positioning performance, which lays the foundation for the study on the compensation methods for main error sources.

4.1 Modeling of Error Sources Within Time Transfer Model

Taking the gravity of eight planets in the solar system and proper motions of pulsars into consideration and rewriting $\boldsymbol{p}_i$ in Eq. (4.1) to be $\boldsymbol{r}_{SC} - \boldsymbol{b}_i$, the time transfer model can be expressed as

$$
\begin{aligned}
d_{SSB} &= t_{SC} + \frac{1}{c}\boldsymbol{n} \cdot \boldsymbol{r}_{SC} \\
&+ \frac{1}{2cD_0}\begin{bmatrix} -\|\boldsymbol{r}_{SC}\|^2 + (\boldsymbol{n} \cdot \boldsymbol{r}_{SC})^2 + 2(\boldsymbol{r}_{SC} \cdot \boldsymbol{V} t_N) \\ -2(\boldsymbol{n} \cdot \boldsymbol{V} t_N)(\boldsymbol{n} \cdot \boldsymbol{r}_{SC}) - 2(\boldsymbol{b} \cdot \boldsymbol{r}_{SC}) + 2(\boldsymbol{n} \cdot \boldsymbol{b})(\boldsymbol{n} \cdot \boldsymbol{r}_{SC}) \end{bmatrix} \\
&+ \sum_{i=1}^{9} \frac{2\mu_i}{c^3} \ln|\boldsymbol{n} \cdot (\boldsymbol{r}_{SC} - \boldsymbol{b}_i) + \|\boldsymbol{r}_{SC} - \boldsymbol{b}_i\|| \qquad (4.1)
\end{aligned}
$$

From Eq. (4.1), error sources influencing the accuracy of the time transfer model mainly include: position errors of the central gravitational body, the Sun and the other celestial bodies, angular position error of the pulsar, the proper motion velocity error of the pulsar, position error of the pulsar and error of the spacecraft-borne atomic clock.

W. Zheng and Y. Wang, *X-ray Pulsar-based Navigation*, Navigation: Science and Technology 5, https://doi.org/10.1007/978-981-15-3293-1_4

4.1.1 Position Error of Central Gravitational Body

Spacecraft move mainly by relying on gravity of the central gravitational body, so its position relative to the SSB can be resolved into the position vector of the spacecraft relative to the central gravitational body and the position vector of the central gravitational body relative to the SSB. Generally speaking, the distance from the spacecraft to the central gravitational body is much shorter than that from the central gravitational body to the SSB. Therefore, the position error of the central gravitational body can significantly affect the accuracy of the time transfer model.

The position of the central gravitational body (a planet) can be predicated by the astronomical Almanac. The currently common ephemeris is the DE405 ephemeris proposed by NASA Jet Propulsion Laboratory (JPL) in 1997. DE405 was established before the upsurge of Mars exploration started with Pathfinder. Hence, the orbit information of the Earth and Mars was obtained based on the distance measurement information of Viking landers from 1976 to 1982 augmented by the radar range observations with an accuracy of about 1 km. Moreover, the orbits of outer planets were determined mainly by optical measurements with accuracies of about tens of kilometers [1]. After DE405, a series of ephemerides such as DE408, DE413 and DE421 have been successively proposed by JPL.

Take an Earth-orbiting spacecraft as an example. By computing partial derivative of Eq. (4.1), the influence function in Eq. (4.2) can be obtained, which illustrates the impact of the position error of the central gravitational body (not the Sun) on the time transfer model.

$$\begin{aligned}\Delta d_{SSB} = {} & \frac{1}{c}\boldsymbol{n}\Delta\boldsymbol{r}_E - \frac{\boldsymbol{r}_E\Delta\boldsymbol{r}_E}{cD_0\|\boldsymbol{r}_{SC}\|} + \frac{(\boldsymbol{n}\cdot\boldsymbol{r}_{SC})(\boldsymbol{n}\cdot\Delta\boldsymbol{r}_E)}{cD_0} \\ & + \frac{\Delta\boldsymbol{r}_E\boldsymbol{V}t_N}{cD_0} - \frac{(\boldsymbol{n}\cdot\boldsymbol{V}t_N)(\boldsymbol{n}\cdot\Delta\boldsymbol{r}_E)}{cD_0} \\ & - \frac{\boldsymbol{b}\Delta\boldsymbol{r}_E}{cD_0} + \frac{(\boldsymbol{n}\cdot\boldsymbol{b})(\boldsymbol{n}\cdot\Delta\boldsymbol{r}_E)}{cD_0} \\ & + \sum_{i=1}^{8}\frac{2\mu_i}{c^3}\frac{\boldsymbol{n}\cdot\Delta\boldsymbol{r}_E + (\boldsymbol{r}_{SC}-\boldsymbol{b}_i)\cdot\Delta\boldsymbol{r}_E/\|\boldsymbol{r}_{SC}-\boldsymbol{b}_i\|}{\boldsymbol{n}\cdot(\boldsymbol{r}_{SC}-\boldsymbol{b}_i)+\|\boldsymbol{r}_{SC}-\boldsymbol{b}_i\|}\end{aligned} \tag{4.2}$$

where $\Delta\boldsymbol{r}_E$ is the error within the position of the Earth.

4.1.2 Position Error of the Sun

For a spacecraft orbiting a planet, the Sun is the main celestial body to cause Shapiro delay. For a deep space explorer orbiting the Sun, the Sun is the central gravitational body. Therefore, the impact caused by the position error of the Sun needs to be analyzed in two cases.

(1) **Spacecraft orbiting a planet**

For a spacecraft orbiting the planet, the position error of the Sun only affects the 3rd and 4th terms on the right hand side of Eq. (4.1). Therefore, the influence function of the position error of the Sun as shown in Eq. (4.3) can be obtained by performing the partial derivative of Eq. (4.1) with respect to the error within position of the Sun.

$$\begin{aligned}\Delta d_{SSB} &= \frac{1}{cD_0}[-\boldsymbol{r}_{SC}\cdot\Delta\boldsymbol{b} + (\boldsymbol{n}\cdot\Delta\boldsymbol{b})(\boldsymbol{n}\cdot\boldsymbol{r}_{SC})] \\ &\quad + \frac{2\mu_S}{c^3}\left[\frac{\boldsymbol{n}\cdot\Delta\boldsymbol{b} + (\boldsymbol{r}_{SC}-\boldsymbol{b})\cdot\Delta\boldsymbol{b}/\|\boldsymbol{r}_{SC}-\boldsymbol{b}\|}{\boldsymbol{n}\cdot(\boldsymbol{r}_{SC}-\boldsymbol{b}) + \|\boldsymbol{r}_{SC}-\boldsymbol{b}\|}\right]\end{aligned} \tag{4.3}$$

where $\Delta\boldsymbol{b}$ is the position error of the Sun.

(2) **Deep space explorer orbiting the Sun**

For a deep space explorer orbiting the Sun, the Sun is the central gravitational body and the position error of the Sun can affect the 2nd, 3rd and 4th terms on the right hand side of Eq. (4.1). In this case, the influence function of the position error of the Sun can be derived and is shown in Eq. (4.4).

$$\begin{aligned}\Delta d_{SSB} &= \frac{1}{c}\boldsymbol{n}\Delta\boldsymbol{b} - \frac{\boldsymbol{b}\Delta\boldsymbol{b}}{cD_0\|\boldsymbol{r}_{SC}\|} + \frac{(\boldsymbol{n}\cdot\boldsymbol{r}_{SC})(\boldsymbol{n}\cdot\Delta\boldsymbol{b})}{cD_0} \\ &\quad + \frac{\Delta\boldsymbol{b}\boldsymbol{V}t_N}{cD_0} - \frac{(\boldsymbol{n}\cdot\boldsymbol{V}t_N)(\boldsymbol{n}\cdot\Delta\boldsymbol{b})}{cD_0} \\ &\quad - \frac{\boldsymbol{b}\Delta\boldsymbol{b} + \Delta\boldsymbol{b}\cdot\boldsymbol{r}_{SC}}{cD_0} + \frac{(\boldsymbol{n}\cdot\Delta\boldsymbol{b})(\boldsymbol{n}\cdot\boldsymbol{r}_{SC}) + (\boldsymbol{n}\cdot\boldsymbol{b})(\boldsymbol{n}\cdot\Delta\boldsymbol{b})}{cD_0} \\ &\quad + \sum_{i=1}^{8}\frac{2\mu_i}{c^3}\frac{\boldsymbol{n}\cdot\Delta\boldsymbol{b} + (\boldsymbol{r}_{SC}-\boldsymbol{b}_i)\cdot\Delta\boldsymbol{b}/\|\boldsymbol{r}_{SC}-\boldsymbol{b}_i\|}{\boldsymbol{n}\cdot(\boldsymbol{r}_{SC}-\boldsymbol{b}_i) + \|\boldsymbol{r}_{SC}-\boldsymbol{b}_i\|}.\end{aligned} \tag{4.4}$$

4.1.3 *Position Error of Other Celestial Bodies*

The influence function of the position error of the Sun as shown in Eq. (4.5) can be obtained by performing the partial derivative of Eq. (4.1) with respect to the error within position of the *i*th celestial body.

$$\Delta d_{SSB} = \frac{2\mu_i}{c^3}\left[\frac{\boldsymbol{n}\cdot\Delta\boldsymbol{b}_i + (\boldsymbol{r}_{SC}-\boldsymbol{b}_i)\cdot\Delta\boldsymbol{b}_i/\|\boldsymbol{r}_{SC}-\boldsymbol{b}_i\|}{\boldsymbol{n}\cdot(\boldsymbol{r}_{SC}-\boldsymbol{b}_i) + \|\boldsymbol{r}_{SC}-\boldsymbol{b}_i\|}\right] \tag{4.5}$$

where $\Delta\boldsymbol{b}_i$ is the position error of the *i*th celestial body.

4.1.4 Angular Position Error of Pulsar

Currently, the angular position of a pulsar is generally determined by VLBI. Based on the current VLBI measurement technology, the accuracy of the pulsar's angular position can reach around 1 mas [2].

From Eq. (4.1), the angular position error pulsar affects the time transfer accuracy by affecting the direction vector of the pulsar. Hence, The influence function of the angular position error of the pulsar as shown in Eq. (4.6) can be obtained by performing the partial derivative of Eq. (4.1) with respect to the error within direction vector of the pulsar.

$$\begin{aligned}\Delta d_{SSB} &= \frac{1}{c}\boldsymbol{r}_{SC}\cdot\Delta\boldsymbol{n} \\ &+ \frac{1}{c}\left[\begin{array}{l}\frac{(\boldsymbol{n}\cdot\boldsymbol{r}_{SC})(\boldsymbol{r}_{SC}\cdot\Delta\boldsymbol{n})}{D_0} - \frac{(\boldsymbol{V}t_N\Delta\boldsymbol{n})(\boldsymbol{n}\cdot\boldsymbol{r}_{SC}) + (\boldsymbol{n}\cdot\boldsymbol{V}t_N)(\boldsymbol{r}_{SC}\cdot\Delta\boldsymbol{n})}{D_0} \\ + \frac{(\boldsymbol{b}\cdot\Delta\boldsymbol{n})(\boldsymbol{n}\cdot\boldsymbol{r}_{SC}) + (\boldsymbol{n}\cdot\boldsymbol{b})(\boldsymbol{r}_{SC}\cdot\Delta\boldsymbol{n})}{D_0}\end{array}\right] \\ &+ \sum_{i=1}^{9}\frac{2\mu_i}{c^3}\frac{\Delta\boldsymbol{n}\cdot(\boldsymbol{r}_{SC}-\boldsymbol{b}_i)}{[\boldsymbol{n}\cdot(\boldsymbol{r}_{SC}-\boldsymbol{b}_i)+\|\boldsymbol{r}_{SC}-\boldsymbol{b}_i\|]}\end{aligned} \tag{4.6}$$

where $\Delta\boldsymbol{n}$ is the error within the direction vector of the pulsar and can be expressed as

$$\Delta\boldsymbol{n} = \begin{bmatrix}-\sin\delta\cos\alpha \\ -\sin\delta\sin\alpha \\ \cos\delta\end{bmatrix}\Delta\delta + \begin{bmatrix}-\cos\delta\sin\alpha \\ \cos\delta\cos\alpha \\ 0\end{bmatrix}\Delta\alpha. \tag{4.7}$$

In Eq. (4.7), α and δ are the right ascension and declination of the pulsar respectively, and $\Delta\alpha$ and $\Delta\delta$ are their errors.

4.1.5 Distance Error of Pulsar

According to the pulsar catalogue prepared by the ATNF, among pulsars observed so far, the nearest one is 0.13 kpc (kiloparsec) away while the farthest one is 30.0 kpc away. As is shown in Eq. (4.1), the delay caused by the pulsar distance for the nearest pulsar and the farthest one are 9.3 μs and 0.04 μs, respectively [3]. Therefore, in order to have a high-precision time transfer performance, the delay caused by the pulsar distance has to be considered. At present, there are mainly two ways to determine the distance of pulsar, including measuring the dispersion and measuring the annual parallax. However, the results of the two methods can differ by a few thousand kpc [4].

Hence, The influence function of the angular position error of the pulsar as shown in Eq. (4.8) can be obtained by performing the partial derivative of Eq. (4.1) with respect to the pulsar distance error.

$$\Delta d_{SSB} = \frac{1}{2cD_0^2}\begin{bmatrix} \|\boldsymbol{r}_{SC}\|^2 \Delta D_0 - (\boldsymbol{n} \cdot \boldsymbol{r}_{SC})^2 \Delta D_0 - 2\boldsymbol{r}_{SC} \cdot \boldsymbol{V} t_N \Delta D_0 \\ +2(\boldsymbol{n} \cdot \boldsymbol{V} t_N)(\boldsymbol{n} \cdot \boldsymbol{r}_{SC}) \Delta D_0 + 2(\boldsymbol{b} \cdot \boldsymbol{r}_{SC}) \Delta D_0 \\ -2(\boldsymbol{n} \cdot \boldsymbol{b})(\boldsymbol{n} \cdot \boldsymbol{r}_{SC}) \Delta D_0 \end{bmatrix} \tag{4.8}$$

where ΔD_0 is the distance error of the pulsar.

4.1.6 Error Within Proper Motion Velocity of Pulsar

Hence, The influence function of the angular position error of the pulsar as shown in Eq. (4.9) can be obtained by performing the partial derivative of Eq. (4.1) with respect to the pulsar distance error.

$$\Delta d_{SSB} = \frac{1}{cD_0}[\boldsymbol{r}_{SC} \cdot \Delta \boldsymbol{V} t_N - (\boldsymbol{n} \cdot \Delta \boldsymbol{V} t_N)(\boldsymbol{n} \cdot \boldsymbol{r}_{SC})] \tag{4.9}$$

where $\Delta \boldsymbol{V}$ is the error within the proper motion velocity of the pulsar.

4.1.7 Error Within Spacecraft-Borne Atomic Clock

From Eq. (4.1), the influence function of the error within the spacecraft-borne atomic clock can be given by

$$\Delta d_{SSB} = \Delta t_{SC} \tag{4.10}$$

where Δt_{SC} is the error of the spacecraft-borne atomic clock.

4.2 Impact of Error Sources

This subsection mainly investigates the impact of the error sources listed above on the time transfer model for the Earth-orbiting spacecraft.

4.2.1 Impact of Error Sources on Time Transfer Model

This subsection will analyze the impact caused by error sources affecting the time transfer model based on 6 rotation-powered pulsars selected by Microcosm Inc. and geosynchronous satellites.

4.2.1.1 Impact Caused by Ephemeris Error of Earth/Sun

(1) Approximation to the relative errors in the planetary ephemerides

A series of ephemerides such as DE408, DE414 and DE421 have been proposed after DE405. As employing new measurement data more accurate than the earlier ones, the later ephemeris is an improvement over earlier ones. The difference between two ephemerides can provide a good approximation to the error within the earlier one.

Figure 4.1 shows the approximated relative errors in the ephemerides of the Earth, Sun and Jupiter over 2012. The ephemeris errors of those planets vary randomly and the ephemeris error of Jupiter is the largest.

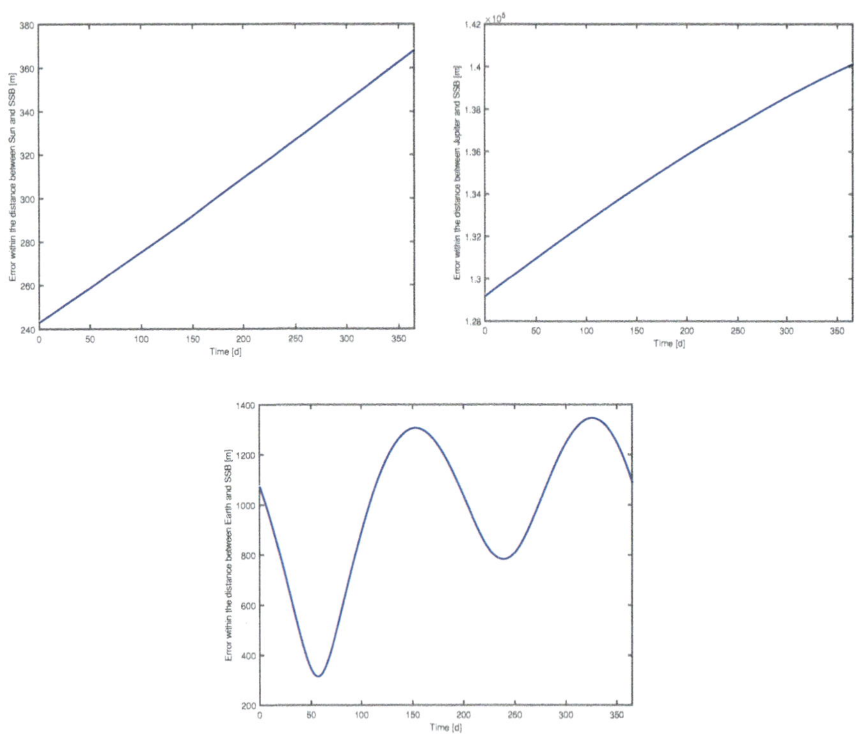

Fig. 4.1 Planet's ephemeris position error approximated by DE421-DE405

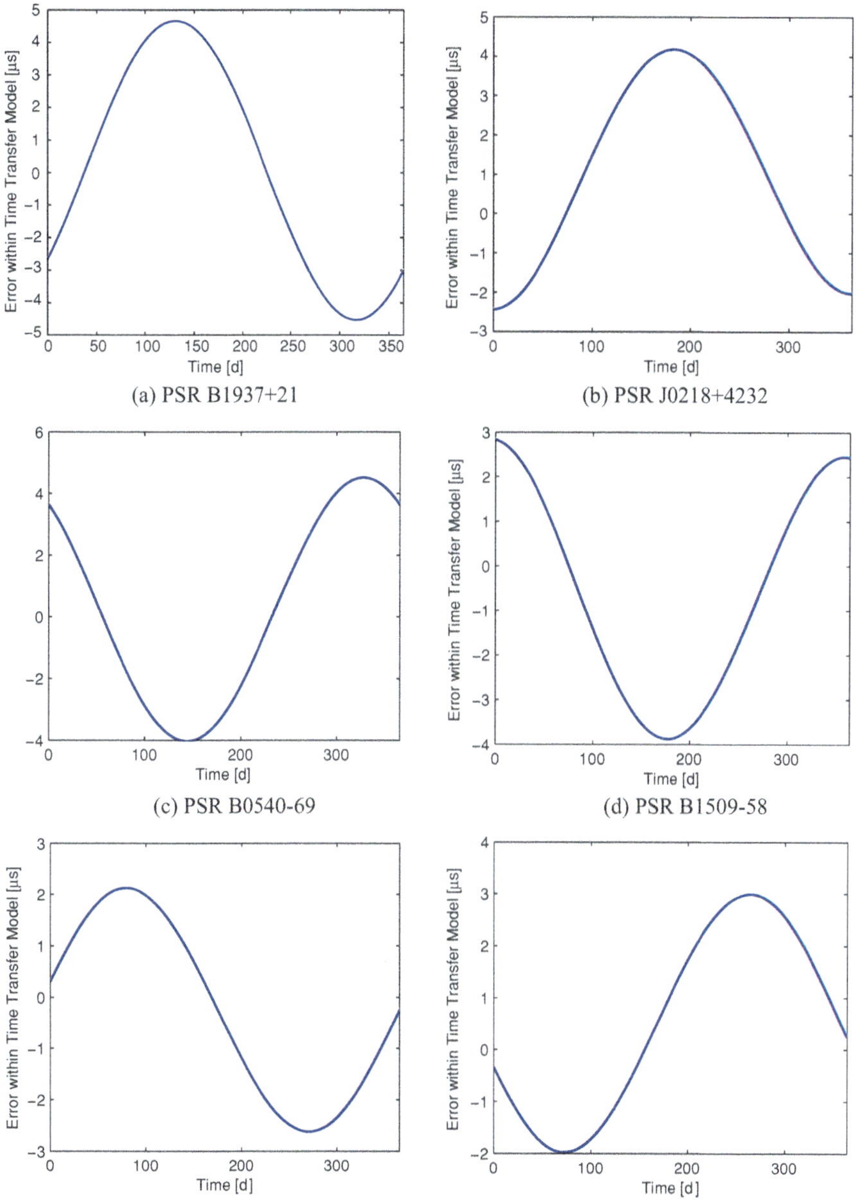

(a) PSR B1937+21

(b) PSR J0218+4232

(c) PSR B0540-69

(d) PSR B1509-58

(e) PSR B1821-24

(f) PSR B0531+21

Fig. 4.2 Impact of Earth ephemeris error on time transfer model

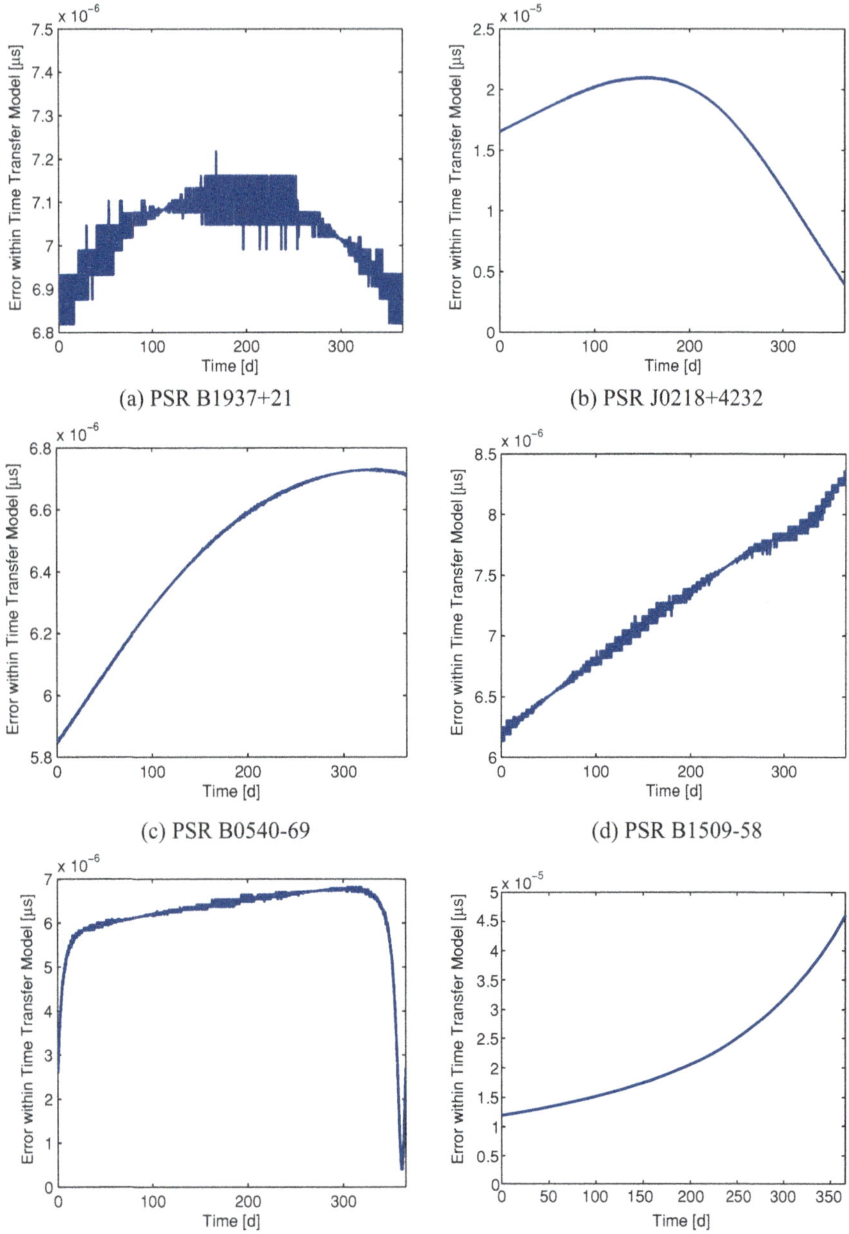

(a) PSR B1937+21

(b) PSR J0218+4232

(c) PSR B0540-69

(d) PSR B1509-58

(e) PSR B1821-24

(f) PSR B0531+21

Fig. 4.3 Impact of Sun ephemeris error on time transfer model

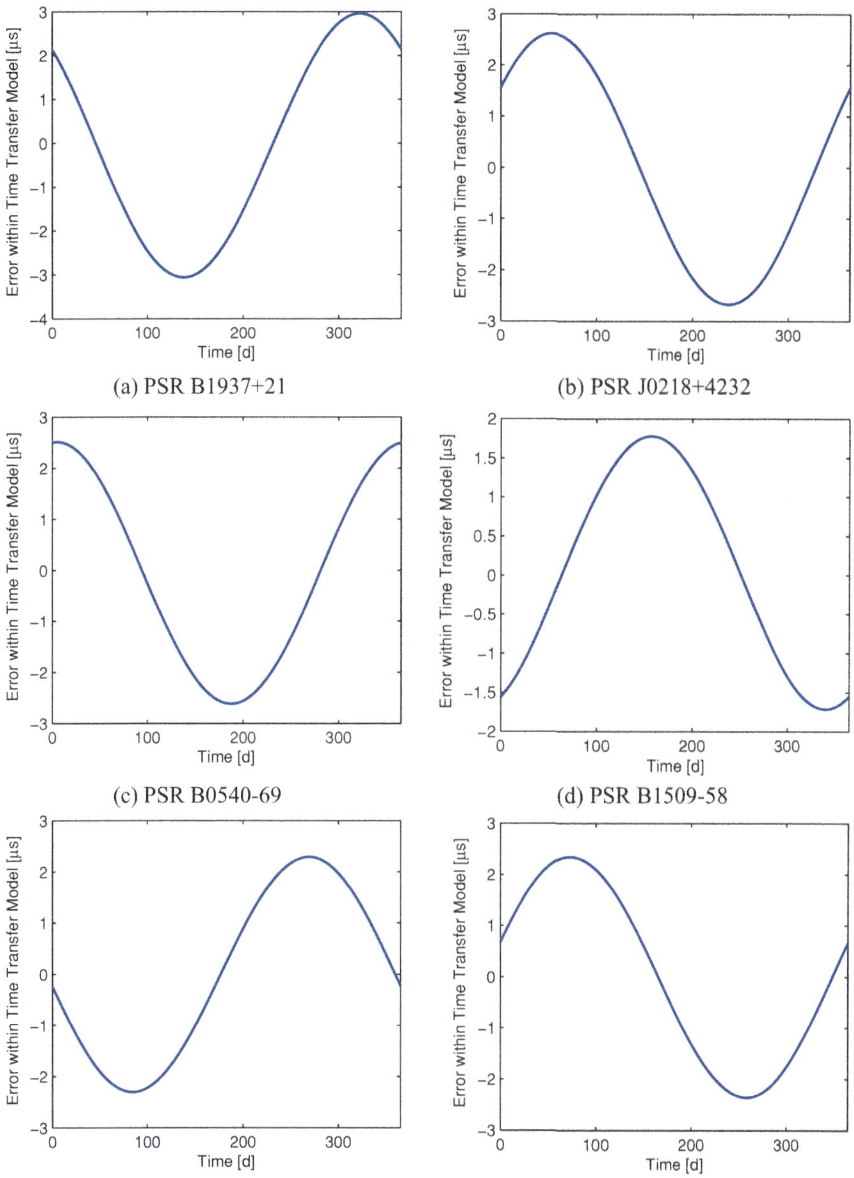

(a) PSR B1937+21

(b) PSR J0218+4232

(c) PSR B0540-69

(d) PSR B1509-58

(e) PSR B1821-24

(f) PSR B0531+21

Fig. 4.4 Impact caused by pulsar's angular position error of 1 mas on time transfer model

(2) **Impact caused by position error of Earth**

The Geosynchronous orbit is selected and the ephemeris DE405 is chosen for providing the position information of the Earth and sun. Differences between DE421 and DE405 are used as the position error of Earth. Figure 4.2 shows the impact of Earth ephemeris error on time transfer model of 6 pulsars in the year 2102.

As shown in Fig. 4.2, the Earth ephemeris error exerts significant impacts on the time transfer model, which cannot reduce as the navigation duration increases.

(3) **Impact caused by position error of sun**

The Geosynchronous orbit is selected and the ephemeris DE405 is chosen for providing the position information of the Earth and Sun. Differences between DE421 and DE405 are used as the position error of the Sun. Figure 4.3 shows the impact of Sun ephemeris error on time transfer model of 6 pulsars in the year 2102.

It can be seen that, the Sun ephemeris error exerts an impact less than 1 ns on the time transfer model throughout the year, and this impact is far slighter than the current accuracy of pulse TOA. Hence, the impact of the Sun ephemeris error can be neglected. As the Sun has the largest mass in the solar system, the ephemeris errors of the other celestial bodies can be neglected in the measurement model.

4.2.1.2 Impact Caused by Angular Position Error of Pulsar

The error of the right ascension and the declination of pulsar is set as 1 mas. Figure 4.4 shows the impact caused by the pulsar's angular position in a whole year.

As Fig. 4.4 shows, these errors vary periodically throughout the year and are mostly greater than 1000 ns.

4.2.1.3 Impact Caused by Distance Error of Pulsar

The distance error of pulsar is set as 30%. Based on DE405, impacts caused by the distance error of 6 pulsars are simulated and the results are shown in Fig. 4.5.

As we can see, impacts caused by the distance error of pulsar are less than 200 ns at most time for the whole year. From Fig. 4.5, we can see that the impact caused by the pulsar distance error on the time transfer model increases as the distance between the pulsar and the Sun decreases. Therefore, this impact needs serious consideration when a pulsar near the Sun is used for navigation.

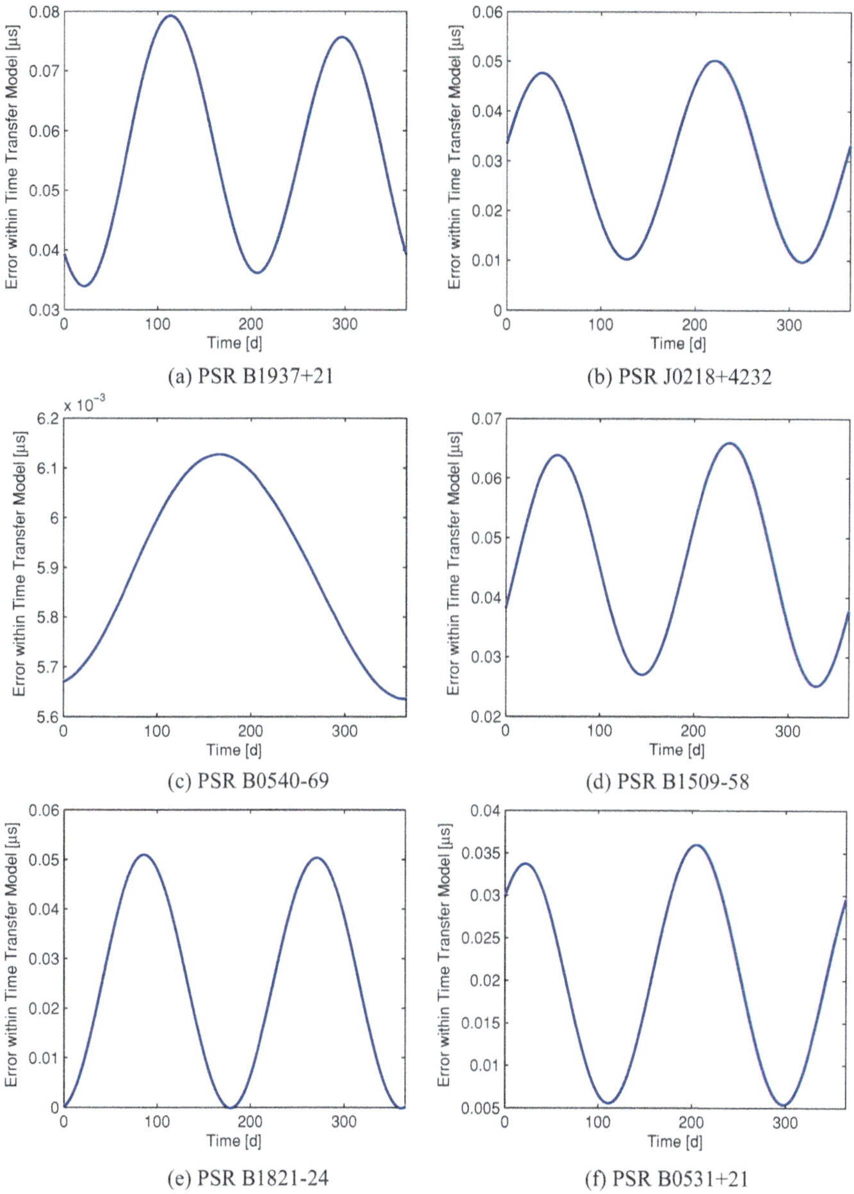

Fig. 4.5 Impact caused by pulsar distance error of 30% on time transfer model

4.2.1.4 Impact Caused by Proper Motion Velocity of Pulsar

The proper motion velocity error of pulsar is taken as 1 mas/year, and the observation period is set as 2 h. Figure 4.6 gives the impact caused by the error within the proper motion velocity of the pulsar on the time transfer model.

As we can see from Fig. 4.6, within the observation duration, the error caused by the pulsar's proper motion velocity error is too small to be shown in the picture. Hence, this error can be neglected.

4.2.2 Impact of Error Source on Template

According to the analysis of Sect. 4.2.1, the Earth ephemeris error and angular position error of pulsar put significant impact on the time transfer model. Thus, this subsection focuses on analyzing the impact of these two error sources on the establishment of the template.

4.2.2.1 Impact of Earth Ephemeris Error

Assuming a spacecraft keeps observing PSR B0531+21 from 2000 to 2014. The Earth ephemeris error is approximated by the difference between DE421 and DE405. Figure 4.7 shows the delay caused by the Earth ephemeris error over the whole observation duration. From Fig. 4.7, the impact of Earth ephemeris error has long-periodical term and short-periodical term.

Figure 4.8 shows the template delay caused by the Earth ephemeris error between 2000 and 2014. The impact of short-periodical term could be greatly reduced in a short duration by employing epoch accumulation, but the impact of long-periodical term has to be reduced by large data accumulation. Thus, the template built within a finite duration is still affected by the Earth ephemeris error, but an appropriate observation strategy could reduce the majority of impact.

4.2.2.2 Impact of Pulsar Angular Position Error

As the influence function of pulsar angular position error is a function with a period of one year, we only need to analyze how the observation data of one year could reduce the impact of pulsar angular position error. Figure 4.9 shows the delay caused by pulsar angular position error of 1 mas over the whole year. The impact of periodical error could be greatly reduced by epoch accumulation, and could be reduced to around 0.1 μs from 1 to 2 μs via the observation of the year. Thus, the impact of periodical error could be reduced by accumulating data according to the period of error. In addition, as shown in Sect. 4.2.1.3, the impact caused by pulsar distance

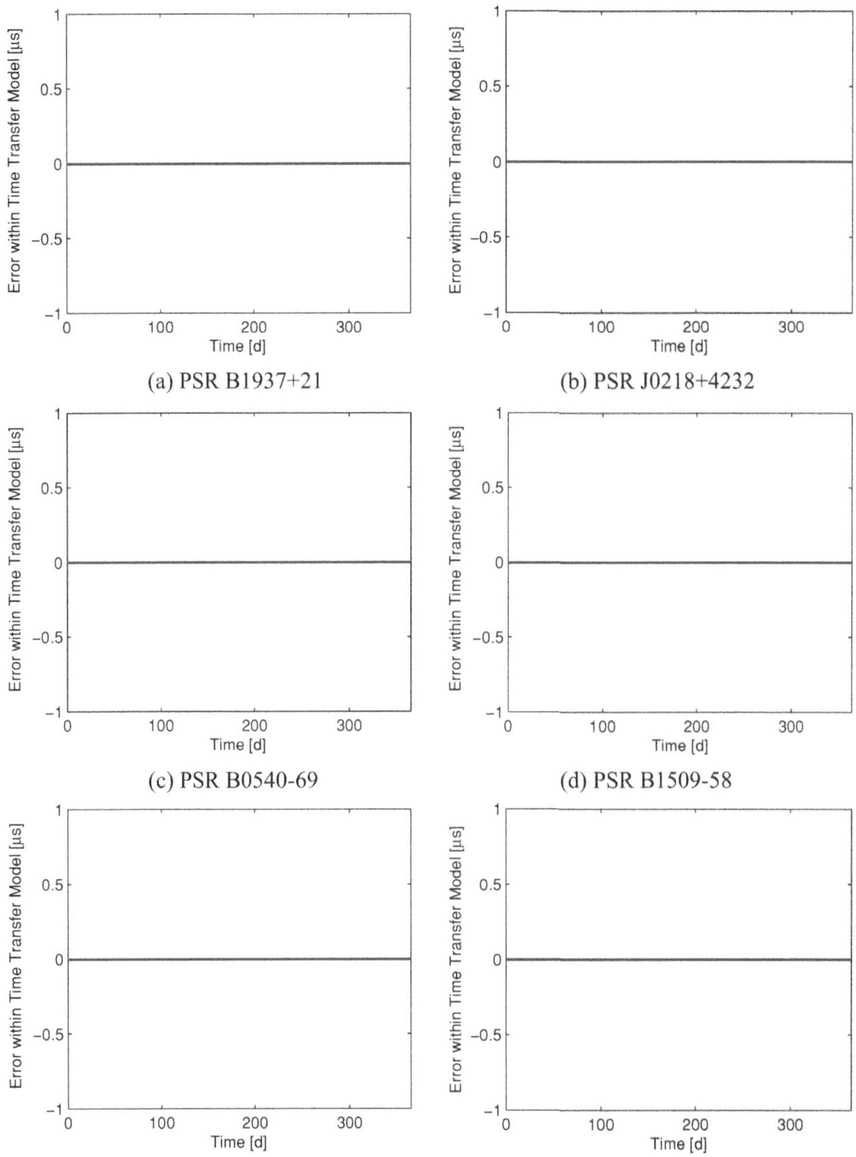

(a) PSR B1937+21

(b) PSR J0218+4232

(c) PSR B0540-69

(d) PSR B1509-58

(e) PSR B1821-24

(f) PSR B0531+21

Fig. 4.6 Impact caused by pulsar's proper motion velocity error of 1 mas/year on time transfer model

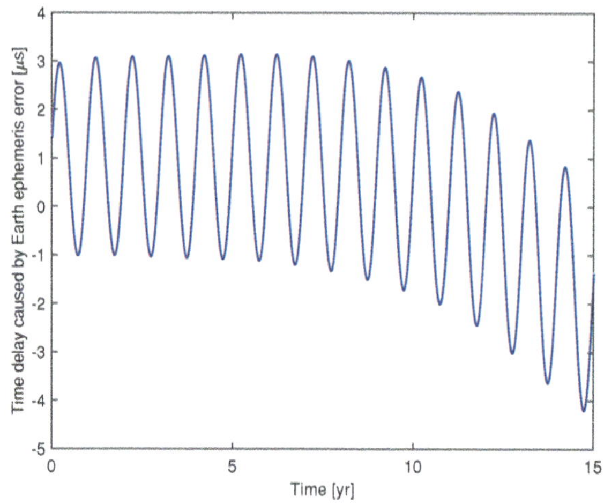

Fig. 4.7 Delay caused by Earth ephemeris error (DE421-DE405) between 2000 and 2014

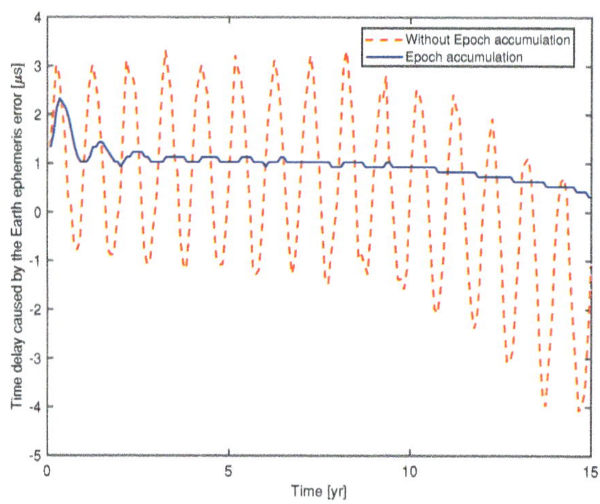

Fig. 4.8 Impact of Earth ephemeris error reduced by epoch accumulation

error is less than 1 μs. Thus, if the accuracy of template is required to be around 1 μs, the impact of pulsar distance error could be neglected.

4.2.3 *Impact of Error Source on Positioning Performance*

In this subsection, the impact of error sources on positioning performance is analyzed by 1000 Monte-Carlo trials.

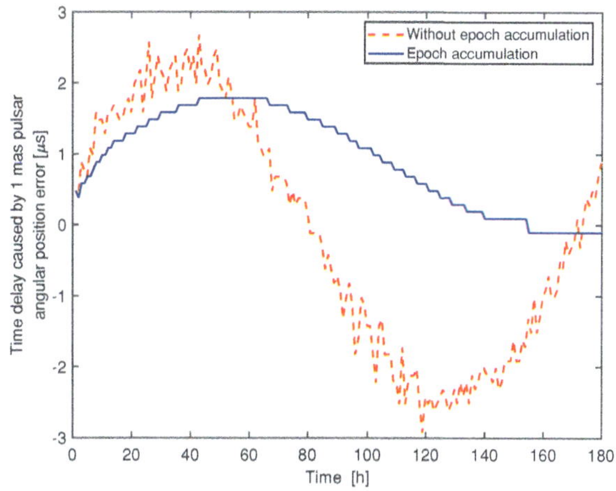

Fig. 4.9 Impact of pulsar angular position error reduced by epoch accumulation

4.2.3.1 Impact Caused by Position Error of Earth

Three pulsars, i.e., PSR B0531+21, PSR B1509−58 and PSR B1821−24, are used for navigation. The observation period is set as 1800 s and the simulation duration is 7 days. Assume that the satellite orbit is geosynchronous orbit and its position error and velocity error are respectively 1 m and 1 m/s. The Earth ephemeris error is taken as 1 km. Figure 4.10 gives the navigation results in this case.

As we can see from Fig. 4.10, despite the presence of Earth ephemeris error, the navigation filter can converge (but not better than 1500 m). It means the Earth ephemeris error lifts up the whole curve of estimation error. Therefore, the impact of the Earth ephemeris error cannot be reduced by prolonging the navigation duration or the other methods applicable to reduce the impact of white Gaussian noise.

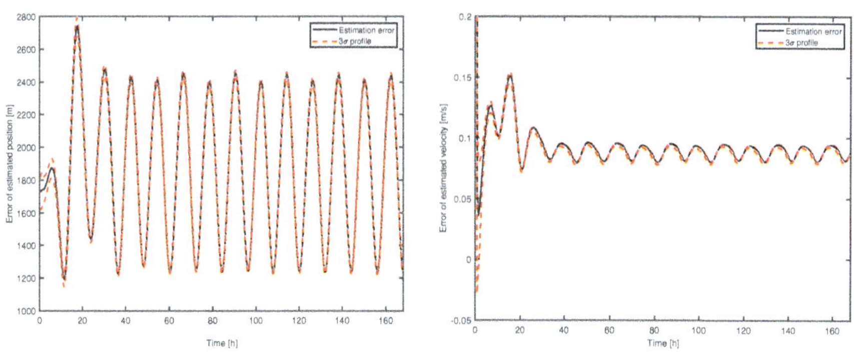

Fig. 4.10 Impact of Earth ephemeris error of 1 km on navigation accuracy

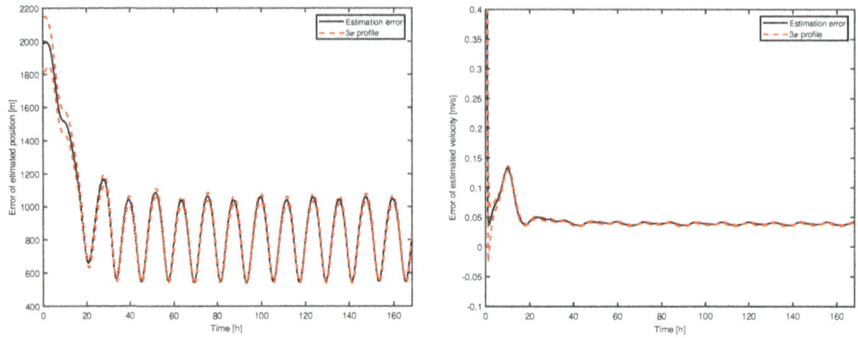

Fig. 4.11 Impact caused by pulsar's angular position error of 1 mas on navigation accuracy

4.2.3.2 Impact Caused by Angular Position Error of Pulsar

Figure 4.11 gives the navigation results for the case where the pulsar angular position error is 1 mas and is obtained by 1000 Monte Carlo simulations.

As Fig. 4.11 shows, the pulsar angular position error of 1 mas can make the mean value of positioning estimation error deviate by about 700 m, swinging between 500 and 800 m. In addition, the mean value of velocity estimation errors deviates by 0.05 m/s.

4.2.3.3 Impact Caused by Distance Error of Pulsar

Figure 4.12 gives the navigation results for the case where the pulsar distance error is 30% and is obtained by 1000 Monte Carlo simulations.

It can be seen that, the pulsar distance error of 30% can make the mean value of positioning estimation error deviate by about 25 m, swinging between 20 and 40 m. In addition, the mean value of velocity estimation errors deviates by 0.005 m/s. Hence,

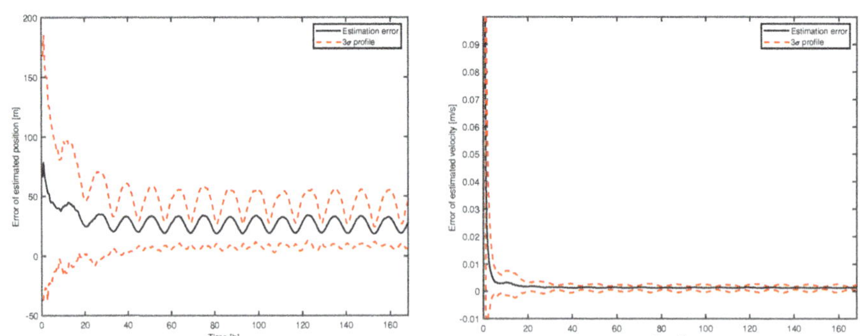

Fig. 4.12 Impact caused by pulsar distance error of 30% on navigation accuracy

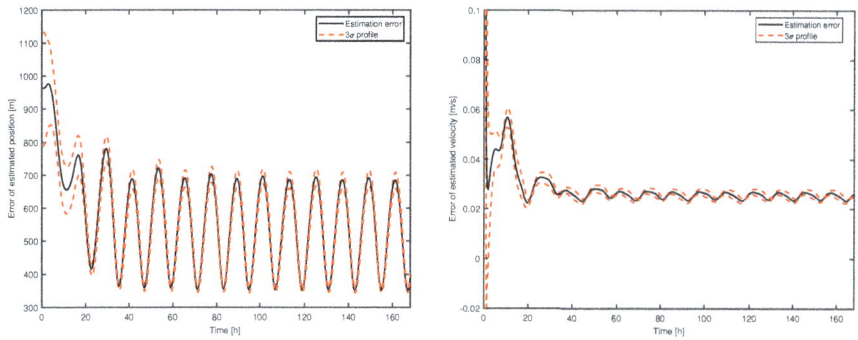

Fig. 4.13 Impact caused by spacecraft-borne atomic clock error of 1 μs on navigation accuracy

the pulsar distance error does not exert great impacts on the positioning performance in most cases, but should be taken into consideration if a high-precision navigation result is needed.

4.2.3.4 Impact of Error Within Spacecraft-Borne Atomic Clock

Figure 4.13 gives the navigation results for the case where the spacecraft-borne atomic clock error is 1 μs and is obtained by 1000 Monte Carlo simulations.

As shown in Fig. 4.13, the spacecraft-borne atomic clock error of 1 μs can cause the mean of positioning estimation error to deviate by 500 m, swinging between 400 and 700 m. Moreover, this error can cause the mean of velocity estimation error to deviate by 0.03 m/s.

4.3 Analysis of Propagation Property of Major Error Sources

4.3.1 Propagation Property of Planet Ephemeris Error

4.3.1.1 Impact of Systematic Biases

Equations (4.2)–(4.5) give the influence function of ephemeris error of celestial bodies. By converting the impact on the time delay to the impact on the distance, we have

$$\Delta d_{SSB} = \frac{1}{c}(B_S + B_M + B_V + B_E + B_{Ma} + B_J + B_{Sat} + B_U + B_N) \tag{4.11}$$

where B_S, B_M, B_V, B_E, B_{Ma}, B_J, B_{Sat}, B_U and B_N are systematic biases resulting respectively from the ephemeris errors of the Sun, Mercury, Venus, the Earth, Mars, Jupiter, Saturn, Uranus and Neptune.

Equation (4.11) shows that the ephemeris errors of Mercury, Venus, Mars, Jupiter, Saturn, Uranus and Neptune could only affect the Shapiro delay. Therefore, the expressions of B_M, B_V, B_{Ma}, B_J, B_{Sat}, B_U and B_N are the same and can be expressed by B_P.

Based on the current status of astronomical observation, we assume that the ephemeris error of Venus, the Earth, Mars or the Sun is 1 km, the orbit deviation of Mercury is 10 km, and the errors of Jupiter, Saturn, Uranus or Neptune are 100 km. Based on the data of DE405 and the parameters in Tables 4.1, 4.2 and 4.3 give the mean values and variances of ephemeris errors-induced systematic biases over 2012.

Table 4.1 Mean value and variance of planet ephemeris errors-induced systematic biases [in m]: part 1

	PSR B1937+21		PSR J0218+4232	
	Mean value	Mean variance	Mean value	Mean variance
B_E	83.8	3.3×10^{-5}	1700.8	1.5×10^{-5}
B_S	9.6×10^{-3}	2.6×10^{-4}	5.0×10^{-3}	1.9×10^{-3}
B_J	1.1×10^{-15}	1.0×10^{-16}	4.0×10^{-15}	1.0×10^{-15}
B_M	2.1×10^{-21}	1.9×10^{-20}	1.9×10^{-20}	3.2×10^{-20}
B_V	9.2×10^{-18}	1.5×10^{-17}	2.7×10^{-17}	3.6×10^{-17}
B_{Ma}	8.7×10^{-19}	3.7×10^{-19}	5.3×10^{-19}	2.8×10^{-19}
B_{Sat}	1.7×10^{-16}	1.1×10^{-17}	6.2×10^{-17}	5.0×10^{-18}
B_U	1.3×10^{-17}	3.7×10^{-19}	1.4×10^{-17}	1.4×10^{-18}
B_N	3.9×10^{-18}	3.0×10^{-19}	5.9×10^{-19}	1.7×10^{-19}

Table 4.2 Mean value and variance of planet ephemeris errors-induced systematic biases [in m]: part 2

	PSR B0540-69		PSR B1509-58	
	Mean value	Mean variance	Mean value	Mean variance
B_E	553.5	2.3×10^{-5}	1582.7	1.7×10^{-5}
B_S	1.3×10^{-2}	1.1×10^{-3}	1.2×10^{-2}	6.1×10^{-4}
B_J	1.2×10^{-15}	8.3×10^{-17}	7.1×10^{-16}	6.5×10^{-17}
B_M	4.5×10^{-21}	1.0×10^{-20}	1.7×10^{-20}	2.2×10^{-20}
B_V	8.3×10^{-18}	1.1×10^{-17}	7.8×10^{-19}	2.0×10^{-17}
B_{Ma}	7.7×10^{-19}	3.4×10^{-19}	1.0×10^{-18}	1.2×10^{-18}
B_{Sat}	1.7×10^{-16}	3.0×10^{-18}	2.6×10^{-16}	3.2×10^{-17}
B_U	9.9×10^{-18}	4.7×10^{-19}	5.9×10^{-18}	3.7×10^{-19}
B_N	1.8×10^{-18}	1.8×10^{-19}	1.9×10^{-18}	2.1×10^{-19}

Table 4.3 Mean value and variance of planet ephemeris errors-induced systematic biases [in m]: part 3

	PSR B1821−24		PSR B0531+21	
	Mean value	Mean variance	Mean value	Mean variance
B_E	1225.6	1.5×10^{-4}	1398.9	1.3×10^{-4}
B_S	1.1×10^{-2}	2.3×10^{-4}	7.1×10^{-3}	6.0×10^{-3}
B_J	1.1×10^{-15}	1.9×10^{-16}	2.0×10^{-14}	2.1×10^{-14}
B_M	4.2×10^{-19}	3.8×10^{-18}	1.1×10^{-19}	7.2×10^{-19}
B_V	6.8×10^{-18}	8.2×10^{-17}	6.7×10^{-17}	2.4×10^{-15}
B_{Ma}	8.2×10^{-17}	8.5×10^{-16}	1.2×10^{-18}	1.6×10^{-18}
B_{Sat}	2.1×10^{-16}	2.8×10^{-17}	1.1×10^{-16}	7.0×10^{-18}
B_U	9.6×10^{-18}	4.0×10^{-19}	9.8×10^{-18}	6.6×10^{-19}
B_N	6.5×10^{-18}	3.1×10^{-19}	3.7×10^{-18}	1.5×10^{-19}

As can be seen from Tables 4.1, 4.2 and 4.3, except PSR B1937+21, the systematic bias induced by the Earth is larger than 500 m, while the systematic bias induced by the Sun is less than 0.1 m. Additionally, the systematic biases caused by the ephemeris error of the other planet are less than 1×10^{-13} m. Therefore, for spacecraft orbiting Earth, the ephemeris error of the Earth is the major error source, which would greatly affect the positioning performance of X-ray pulsar-based navigation system.

4.3.1.2 State Model of Systematic Bias

Generally speaking, the Earth ephemeris error is produced by the perturbation arising from asteroids. Asteroids are mainly in the asteroid belt which is between Mars and Jupiter. While the orbits of significant asteroids are known with sufficient accuracy, most of their masses are not. A mass ring was proposed to account for all asteroids. In this method, $M_{ring} \approx (5 \pm 1) \times 10^{-10}\, M_S$ (M_S is the solar mass), the mean mass ring radius $R_{ring} \approx 2.80$ AU (1 AU $= 1.496 \times 10^{11}$ m), with 3% uncertainty. Define $\delta \boldsymbol{r}$ to be the difference between the original system and the system with the mass ring. $\delta \boldsymbol{r}$ can be integrated by

$$\delta \ddot{\boldsymbol{r}} = -\mu_S \left[\frac{\delta \boldsymbol{r}}{\|\boldsymbol{r}\|^3} - \frac{\boldsymbol{r}(\boldsymbol{r} \cdot \delta \boldsymbol{r})}{\|\boldsymbol{r}\|^5} \right] + GM_{ring} \left[\frac{\boldsymbol{r}_r}{\|\boldsymbol{r}_r\|^3} - \frac{\boldsymbol{R}_{ring}}{\|\boldsymbol{R}_{ring}\|^5} \right] \tag{4.12}$$

where $\boldsymbol{r}$ is the position of the Earth relative to the Sun, $\boldsymbol{r}_r$ is the position of the Earth relative to the mass ring, and $\boldsymbol{R}_{ring}$ is the position of the mass ring relative to the Sun. Although the impacts of the ephemeris error of the other celestial bodies are not taken into account in Eq. (4.12), their impacts can be estimated effectively. Considering that the mass of the solar system is mainly composed of the Sun and Jupiter, the perturbation acceleration induced by the Jupiter ephemeris error is the maximum, which is given by

$$\delta \boldsymbol{a}_J = -\mu_J \left[\left(\frac{\boldsymbol{r}_J + \delta \boldsymbol{r}_J}{\|\boldsymbol{r}_J + \delta \boldsymbol{r}_J\|^3} - \frac{\boldsymbol{r}_J}{\|\boldsymbol{r}_J\|^3} \right) - \left(\frac{\boldsymbol{r}_{Jr} + \delta \boldsymbol{r}_J}{\|\boldsymbol{r}_{Jr} + \delta \boldsymbol{r}_J\|^3} - \frac{\boldsymbol{r}_{Jr}}{\|\boldsymbol{r}_{Jr}\|^3} \right) \right] \tag{4.13}$$

where μ_J is the gravitational constant of Jupiter, $\delta \boldsymbol{r}_J$ is the position deviation of Jupiter relative to the Sun, $\boldsymbol{r}_J$ is the position of Jupiter relative to the Sun, and $\boldsymbol{r}_{Jr}$ is the position of the Earth relative to Jupiter. When Jupiter coincides with the Earth, the distance between the Earth and Jupiter is the minimum (about 6.28×10^8 km). In this case, assuming that $\delta \boldsymbol{r}$ and $\delta \boldsymbol{r}_J$ are [1 km, 1 km, 1 km] and [100 km, 100 km, 100 km] respectively, δa_J is about 2.73×10^{-22} m/s^2. Now the perturbation acceleration of asteroids is about 8.2×10^{-13} m/s^2. Therefore, it is reasonable to neglect the ephemeris error of the other celestial bodies in Eq. (4.12).

Noting that Eq. (4.2) is valid in the Sun-centered inertial system, $\delta \boldsymbol{r}$ can be written as

$$\delta \boldsymbol{r} = \Delta \boldsymbol{r}_E - \Delta \boldsymbol{b}. \tag{4.14}$$

Assuming that SSB is always at the origin of the inertial reference system, the position of the Sun relative to SSB can be determined by

$$M_S \boldsymbol{b} + \sum_{k=1}^{n} M_k \boldsymbol{r}_k = \boldsymbol{0} \tag{4.15}$$

where M_k ($k = 1, 2, \ldots, n$) is the kth celestial body with known position and mass, and $\boldsymbol{r}_k$ is the position of such celestial body relative to SSB. Considering $\boldsymbol{r}_k = \boldsymbol{R}_k + \boldsymbol{b}$ ($\boldsymbol{R}_k$ is the position of the kth celestial body relative to the Sun), Eq. (4.15) can be rewritten as

$$\left(M_S + \sum_{k=1}^{n} M_k \right) \boldsymbol{b} + \sum_{k=1}^{n} M_k \boldsymbol{R}_k = \boldsymbol{0}. \tag{4.16}$$

Assuming the position deviation of the Sun is also caused by the asteroid with unknown mass yields

$$\left(M_S + \sum_{k=1}^{n} M_k \right) (\boldsymbol{b} + \Delta \boldsymbol{b}) + \sum_{k=1}^{n} M_k \boldsymbol{R}_k + \sum_{k=1}^{n} M_k \delta \boldsymbol{R}_k + M_{ring} \boldsymbol{R}_{ring} = \boldsymbol{0} \tag{4.17}$$

where $\delta \boldsymbol{R}_k$ is the position error of the kth celestial body, which can be solved by Eq. (4.12). From Eqs. (4.16) and (4.17), we have

$$\left(M_S + \sum_{k=1}^{n} M_k\right)\Delta\boldsymbol{b} + \sum_{k=1}^{n} M_k\delta\boldsymbol{R}_k + M_{ring}\boldsymbol{R}_{ring} = \boldsymbol{0}. \tag{4.18}$$

Hence,

$$\Delta\boldsymbol{b} = -\frac{1}{M_S + \sum_{k=1}^{n} M_k}\left[\sum_{k=1}^{n} M_k\delta\boldsymbol{R}_k + M_{ring}\boldsymbol{R}_{ring}\right]. \tag{4.19}$$

Equation (4.19) can be simplified further into

$$\Delta\boldsymbol{b} = -\frac{1}{M_{SS}}\left[\sum_{k=1}^{n} M_k\delta\boldsymbol{R}_k + M_{ring}\boldsymbol{R}_{ring}\right] \tag{4.20}$$

where M_{SS} is the mass of solar system. M_{SS} can be simplified as M_S if the corresponding approximated error can be accepted.

In order to solve Eq. (4.19), the perturbation of eight planets caused by asteroids should be considered. Based on the positions of the Earth and the Sun obtained by DE405, assuming that initial values of $\delta\boldsymbol{r}$ are [1 km, 1 km, 1 km], the increment of position deviation within 1800 s can be obtained by integrating Eq. (4.12). Now, $\Delta\boldsymbol{r}_E$ can be determined as about 2.3×10^{-4} m by using Eqs. (4.14) and (4.19). In addition, the increment of $\Delta\boldsymbol{r}_E$ in one day is less than 0.5 m.

Therefore, the systematic bias induced by the ephemeris error of the Earth can be established as a slow time-varying process with a state model of

$$\dot{B}_E = 0 + w_E \tag{4.21}$$

where w_E is the process noise.

4.3.2 Propagation Property of Pulsar Angular Position Error

As the spacecraft orbiting Earth is about 1 AU from the Sun, the second and third terms in Eq. (4.6) are less than 4 ns and can be neglected. Converting the impact on the time transfer model into the impact on the measuring distance, we have

$$\Delta d_{SSB} = \frac{1}{c}B_{PN}. \tag{4.22}$$

B_{PN} can be simplified as

$$B_{PN} = \boldsymbol{r}_{SC} \cdot \Delta\boldsymbol{n}. \tag{4.23}$$

Considering $\boldsymbol{r}_{SC/E} \ll \boldsymbol{r}_B$, B_{PN} can be expressed as

$$B_{PN} = \boldsymbol{r}_E \cdot \Delta\boldsymbol{n}. \tag{4.24}$$

Hence, the time varying rate of B_{PN} is

$$\dot{B}_{PN} = \dot{\boldsymbol{r}}_E \cdot \Delta\boldsymbol{n}. \tag{4.25}$$

Now the absolute value of the time varying rate of B_{PN} is

$$\left|\dot{B}_{PN}\right| = |\dot{\boldsymbol{r}}_E \cdot \Delta\boldsymbol{n}| \leq \|\Delta\boldsymbol{n}\| \|\dot{\boldsymbol{r}}_E\| \tag{4.26}$$

where $\|\dot{\boldsymbol{r}}_E\|$ is the velocity of the Earth. When the Earth arrives at the aphelion, its velocity reaches the maximum (about 30.3 km/s).

From Eq. (4.7), we know

$$\|\Delta\boldsymbol{n}\| = \sqrt{(\cos\delta)^2(\Delta\alpha)^2 + (\Delta\delta)^2} \leq \sqrt{(\Delta\alpha)^2 + (\Delta\delta)^2}. \tag{4.27}$$

Then, Eq. (4.26) can be further transformed to be

$$\left|\dot{B}_{PN}\right| \leq \|\Delta\boldsymbol{n}\| \|\dot{\boldsymbol{r}}_E\| \leq \sqrt{(\Delta\alpha)^2 + (\Delta\delta)^2} \cdot \max(\dot{\boldsymbol{r}}_E). \tag{4.28}$$

Assuming that the pulsar angular position error is 1 mas, $\left|\dot{B}_{PN}\right|$ is less than 2.0775×10^{-4} m/s. In other words, the variation of B_{PN} in one day is less than 18 m. Therefore, if the observation duration of a pulsar is shorter than 1 h, the variation of B_{PN} is less than 1 m. Consequently, B_{PN} can be also established as a slow time-varying process with a state model of

$$\dot{B}_{PN} = 0 + w_{PN} \tag{4.29}$$

where w_{PN} is the process noise.

4.3.3 *Propagation Property of Pulsar Distance Error*

Rewrite the influence function in Eq. (4.8) as the impact on measuring distance, namely

$$\Delta d_{SSB} = \frac{1}{c} B_D \tag{4.30}$$

Neglecting the proper motion of pulsar, the derivative of B_D with respect to time is given by

$$\dot{B}_D = \frac{\Delta D_0}{D_0^2}[\boldsymbol{r}_{SC} \cdot \dot{\boldsymbol{r}}_{SC} - (\boldsymbol{n} \cdot \boldsymbol{r}_{SC})(\boldsymbol{n} \cdot \dot{\boldsymbol{r}}_{SC}) + \boldsymbol{b} \cdot \dot{\boldsymbol{r}}_{SC} - (\boldsymbol{n} \cdot \boldsymbol{b})(\boldsymbol{n} \cdot \dot{\boldsymbol{r}}_{SC})]. \tag{4.31}$$

Hence

$$\begin{aligned}\dot{B}_D &\le \frac{\Delta D_0}{D_0^2}(\|\boldsymbol{r}_{SC}\| \cdot \|\dot{\boldsymbol{r}}_{SC}\| + \|\boldsymbol{n} \cdot \boldsymbol{r}_{SC}\|\|\boldsymbol{n} \cdot \dot{\boldsymbol{r}}_{SC}\| + \|\boldsymbol{b} \cdot \dot{\boldsymbol{r}}_{SC}\| + \|\boldsymbol{n} \cdot \boldsymbol{b}\|\|\boldsymbol{n} \cdot \dot{\boldsymbol{r}}_{SC}\|) \\ &\le \frac{2\Delta D_0}{D_0^2}(\|\boldsymbol{r}_{SC}\| + \|\boldsymbol{b}\|)\|\dot{\boldsymbol{r}}_{SC}\|, \end{aligned} \tag{4.32}$$

considering $\boldsymbol{r}_{\mathrm{SC/E}} \ll \boldsymbol{r}_{\mathrm{B}}$, $\boldsymbol{r}_{SC} \approx \boldsymbol{r}_E$. It is obvious that when $\boldsymbol{r}_E$, $\boldsymbol{b}$ and $\dot{\boldsymbol{r}}_E$ reach the maximum and D_0 reaches the minimum, Eq. (4.32) has the maximum value.

As can be seen from the data generated from DE405 ephemeris, the maximum position and velocity of the Earth are 1.523×10^8 km and 30.2805 km/s respectively, and the maximum value of the Sun position is 5.2428×10^5 km. It can be learned that the minimum distance amongst these pulsars is 2 kpc. Assume the uncertainty of distance is 30%, then $\dot{B}_D \le 9.2578 \times 10^{-5}$ m. Thus, B_D can be also modeled as a slow-time varying process with a state model of

$$\dot{B}_D = 0 + w_D \tag{4.33}$$

where w_D is the process noise.

4.3.4 Propagation Property of Clock Error of Spacecraft-Borne Atomic Clock

Both clock errors of atomic clock and crystal oscillator comprise the systematic error and random error

$$\delta t = \delta t_1 + \delta t_2 \tag{4.34}$$

where δt is the clock error, δt_1 is the systematic error, and δt_2 is the random error.

The systematic error can be expressed as

$$\delta t_1 = a_0 + a_1 \cdot (t - t_0) + a_2 \cdot (t - t_0)^2 \tag{4.35}$$

where a_0 is the constant error of clock error, a_1 is the error within the clock drift, a_2 is the clock frequency drift, t_0 is reference time, and t is the target time.

The random error, δt_2, is a complicated random process and is usually viewed to be the sum of five independent random processes (Table 4.4).

The total spectral density is

Table 4.4 Spectrum densities and Allan variances of the five random process

No.	Random error	Spectral density	Allan variance
1	Frequency random walk	$h_{-2} \cdot f^{-2}$	$\sigma_{-2}^2(\tau) = h_{-2} \cdot \frac{(2\pi)^2}{6}\lvert\tau\rvert$
2	Frequency flicker	$h_{-1} \cdot f^{-1}$	$\sigma_{-1}^2(\tau) = h_{-1} \cdot 2\ln 2$
3	Frequency white noise	h_0	$\sigma_0^2(\tau) = h_0 \cdot \frac{1}{2\lvert\tau\rvert}$
4	Phase scintillation	$h_1 \cdot f$	$\sigma_1^2(\tau) =$ $h_1 \cdot \frac{1}{(2\pi\tau)^2}\{3[2+\ln(2\pi f_n\lvert\tau\rvert)] - \ln 2\}$
5	Phase white noise	$h_2 \cdot f^2$	$\sigma_2^2(\tau) = h_2 \cdot \frac{3f_n}{(2\pi\tau)^2}$

Note f_n is the upper limit of clock frequency, and $h_{-2}, h_{-1}, h_0, h_1, h_2$ are the clock error parameters

$$S_y(f) = h_{-2} \cdot f^{-2} + h_{-1} \cdot f^{-1} + h_0 + h_1 \cdot f + h_2 \cdot f^2. \tag{4.36}$$

The total Allan variance is

$$\sigma_y^2(\tau) = \sigma_{-2}^2(\tau) + \sigma_{-1}^2(\tau) + \sigma_0^2(\tau) + \sigma_1^2(\tau) + \sigma_2^2(\tau) \tag{4.37}$$

where the Allan variance is defined by

$$\sigma_y^2(\tau) \triangleq \lim_{n\to\infty} \frac{1}{2(n-1)} \sum_{i=1}^{n-1} (y_{i+1} - y_i)^2 \tag{4.38}$$

or

$$\sigma_y^2(\tau) \triangleq \lim_{n\to\infty} \frac{1}{2(n-2)\tau^2} \sum_{i=1}^{n-2} (x_{i+2} - 2x_{i+1} + x_i)^2 \tag{4.39}$$

where n is the sampling quantity, τ is the sampling interval, y_i is the clock frequency relative deviation measured at the ith sampling, and x_i is the phase error measured at the ith sampling.

a_0, a_1, a_2, t_0 can be fitted by employing the collected clock error data. And then, the clock error only contains the random error. $h_{-2}, h_{-1}, h_0, h_1, h_2$ can be fitted by using Eq. (4.40) on the premise that a series of $\sigma_y^2(\tau)$ calculating at a series of sampling intervals have been obtained.

$$\begin{aligned}\sigma_y^2(\tau) &= h_{-2} \cdot \frac{(2\pi)^2}{6}\lvert\tau\rvert + h_{-1} \cdot 2\ln 2 + h_0 \cdot \frac{1}{2\lvert\tau\rvert} \\ &+ h_1 \cdot \frac{1}{(2\pi\tau)^2}\{3[2+\ln(2\pi f_n\lvert\tau\rvert)] - \ln 2\} + h_2 \cdot \frac{3f_n}{(2\pi\tau)^2}.\end{aligned} \tag{4.40}$$

For spacecraft-borne atomic clock, a_2 can be neglected, and a_1 is normally less than 1×10^{-11} Hz. In other words, clock error would only increase by less than 18 ns when the observation period of pulsar is 1800 s. Therefore, the impact of clock error can be regarded as a slow time-varying process. Let $B_T = c\Delta t_{SC}$, and we have the state model of clock error as

$$\dot{B}_T = 0 + w_T \tag{4.41}$$

with w_T of the process noise.

4.4 Systematic Biases Compensation Method Based on Augmented State

According to the analysis in Sect. 5.3, for Earth orbiting spacecraft, the Earth position error, the pulsar angular position error, the pulsar distance error and the pulsar spacecraft-borne atomic clock error all can be modeled as a time slow-varying process. Thus, the comprehensive impact of all the systematic biases can also be established as a time slow-varying process with a state model of the form

$$\dot{B} = 0 + w \tag{4.42}$$

where B is the comprehensive impact of the systematic biases, and w is the process noise.

The augmented state method is a straightforward method to compensate the comprehensive impact of the systematic biases. The basic principle of the augmented state method is that the systematic bias is augmented to be the components of the state waiting to be estimated during the navigation process.

4.4.1 Navigation System

4.4.1.1 State Model

Augmenting B in Eq. (4.42) yields an augmented state model of the form

$$\begin{bmatrix} \dot{\boldsymbol{r}} \\ \dot{\boldsymbol{v}} \\ \dot{\boldsymbol{B}} \end{bmatrix} = \begin{bmatrix} \boldsymbol{v} \\ \boldsymbol{a} \\ \boldsymbol{0} \end{bmatrix} + \begin{bmatrix} \boldsymbol{w}_r \\ \boldsymbol{w}_v \\ \boldsymbol{w}_B \end{bmatrix} \tag{4.43}$$

where $\boldsymbol{X} = [\boldsymbol{r}^{\mathrm{T}}, \boldsymbol{v}^{\mathrm{T}}, \boldsymbol{B}^{\mathrm{T}}]^{\mathrm{T}}$ is the augmented state vector, and $\boldsymbol{B} = [B_1, B_2, \ldots, B_i, \ldots, B_N]^{\mathrm{T}}$ with B_i of the systematic bias in the measurement of ith pulsar. $\boldsymbol{W} = [\boldsymbol{w}_r^{\mathrm{T}}, \boldsymbol{w}_v^{\mathrm{T}},$

w_B^T] is the augmented state process noise modeled as a zero-mean Gaussian white noise with the covariance of $\boldsymbol{Q}$.

4.4.1.2 Measurement Model

Considering the systematic bias, the time transfer model for the measurement of ith pulsar is given by

$$\begin{aligned} d_{SSB}^i &= t_S^i + \frac{1}{c}\tilde{\boldsymbol{n}}^i \cdot (\boldsymbol{r} + \tilde{\boldsymbol{r}}_E) \\ &+ \frac{1}{2c\tilde{D}_0^i}\left\{\begin{array}{l} \left[\tilde{\boldsymbol{n}}^i \cdot (\boldsymbol{r} + \tilde{\boldsymbol{r}}_E)\right]^2 - \|\boldsymbol{r} + \tilde{\boldsymbol{r}}_E\|^2 + 2\left[(\boldsymbol{r} + \tilde{\boldsymbol{r}}_E) \cdot \tilde{\boldsymbol{V}}^i t_N\right] \\ -2\left(\tilde{\boldsymbol{n}}^i \cdot \tilde{\boldsymbol{V}}^i t_N\right)\left[\tilde{\boldsymbol{n}}^i \cdot (\boldsymbol{r} + \tilde{\boldsymbol{r}}_E)\right] \\ -2\left[\boldsymbol{b} \cdot (\boldsymbol{r} + \tilde{\boldsymbol{r}}_E)\right] + 2\left(\tilde{\boldsymbol{n}}^i \cdot \boldsymbol{b}\right)\left[\tilde{\boldsymbol{n}}^i \cdot (\boldsymbol{r} + \tilde{\boldsymbol{r}}_E)\right] \end{array}\right\} \\ &+ \sum_{j=1}^{9} \frac{2\mu_i}{c^3} \ln\left|\frac{\tilde{\boldsymbol{n}}^i \cdot (\boldsymbol{r} + \tilde{\boldsymbol{r}}_E - \boldsymbol{b}_j) + \|\boldsymbol{r} + \tilde{\boldsymbol{r}}_E - \boldsymbol{b}_j\|}{\tilde{\boldsymbol{n}}^i \cdot \boldsymbol{b}_j + \|\boldsymbol{b}_j\|}\right| + \frac{1}{c}B \end{aligned} \tag{4.44}$$

Assume the measured value $\boldsymbol{Z}$= $[c(d_{SSB}^1 - t_S^1), \ldots, c(d_{SSB}^i - t_S^i), \ldots, c(d_{SSB}^N - t_S^N)]^T$, the measured noise $\boldsymbol{V}$= $[v^1, \ldots, v^i, \ldots, v^N]^T$ and the system bias $\boldsymbol{B}$= $[B^1, \ldots, B^i, \ldots, B^N]^T$, where $1 \leq i \leq N$, N is the number of observed pulsars. The measurement model is

$$\boldsymbol{Z} = \boldsymbol{h}(\boldsymbol{X}) + \boldsymbol{v} \tag{4.45}$$

where $\boldsymbol{h}(\boldsymbol{X})$= $[\boldsymbol{h}_p^1(\boldsymbol{X}), \ldots, \boldsymbol{h}_p^i(\boldsymbol{X}), \ldots, \boldsymbol{h}_p^N(\boldsymbol{X})]^T$ and $\boldsymbol{h}_p^i(\boldsymbol{X})$ is the measurement model observing ith pulsar with the expression of

$$\begin{aligned} \boldsymbol{h}_P^i(\boldsymbol{X}) &= \tilde{\boldsymbol{n}}^i \cdot (\boldsymbol{r} + \tilde{\boldsymbol{r}}_E) \\ &+ \frac{1}{2\tilde{D}_0^i}\left\{\begin{array}{l} \left[\tilde{\boldsymbol{n}}^i \cdot (\boldsymbol{r} + \tilde{\boldsymbol{r}}_E)\right]^2 - \|\boldsymbol{r} + \tilde{\boldsymbol{r}}_E\|^2 + 2\left[(\boldsymbol{r} + \tilde{\boldsymbol{r}}_E) \cdot \tilde{\boldsymbol{V}}^i t_N\right] \\ -2\left(\tilde{\boldsymbol{n}}^i \cdot \tilde{\boldsymbol{V}}^i t_N\right)\left[\tilde{\boldsymbol{n}}^i \cdot (\boldsymbol{r} + \tilde{\boldsymbol{r}}_E)\right] - 2\left[\boldsymbol{b} \cdot (\boldsymbol{r} + \tilde{\boldsymbol{r}}_E)\right] + 2\left(\tilde{\boldsymbol{n}}^i \cdot \boldsymbol{b}\right)\left[\tilde{\boldsymbol{n}}^i \cdot (\boldsymbol{r} + \tilde{\boldsymbol{r}}_E)\right] \end{array}\right\} \\ &+ \sum_{j=1}^{9} \frac{2\mu_i}{c^2} \ln\left|\frac{\tilde{\boldsymbol{n}}^i \cdot (\boldsymbol{r} + \tilde{\boldsymbol{r}}_E - \boldsymbol{b}_j) + \|\boldsymbol{r} + \tilde{\boldsymbol{r}}_E - \boldsymbol{b}_j\|}{\tilde{\boldsymbol{n}}^i \cdot \boldsymbol{b}_j + \|\boldsymbol{b}_j\|}\right|. \end{aligned} \tag{4.46}$$

4.4.2 *Observability Analysis*

Observability analysis should be performed in advance to verify the effectiveness of augmented state model, as the complete observability is the necessary condition for ensuring a navigation system could achieve a desired estimation performance.

4.4.2.1 Linearized Observability Criterion for Nonlinear Systems

The linearized observability criterion is the most common observability analysis method for practical applications.

For the following nonlinear system

$$\begin{cases} \dot{x} = f(x) \\ y = h(x) \end{cases}, \tag{4.47}$$

let $\boldsymbol{x}^* \in \boldsymbol{R}^n$ meet the condition

$$f(\boldsymbol{x}*) = \boldsymbol{0}_n, \boldsymbol{h}(\boldsymbol{x}*) = \boldsymbol{0}_m \tag{4.48}$$

i.e. $\boldsymbol{x}^*$ is the balance point of f.

And then, a linearized system around $\boldsymbol{x}^*$ can be obtained by linearizing Eq. (4.47) around $\boldsymbol{x}^*$ and is of the form

$$\begin{cases} \dot{x} = \boldsymbol{A}\boldsymbol{x} \\ y = \boldsymbol{C}\boldsymbol{x} \end{cases} \tag{4.49}$$

where

$$\boldsymbol{A} = \left.\frac{\partial \boldsymbol{f}(\boldsymbol{x})}{\partial \boldsymbol{x}}\right|_{x=x^*}, \boldsymbol{C} = \left.\frac{\partial \boldsymbol{h}(\boldsymbol{x})}{\partial \boldsymbol{x}}\right|_{x=x^*}. \tag{4.50}$$

Lee and Markus gave the following Lemma:

If the system in Eq. (4.49) is completely observable, there is a $\sigma > 0$, which makes the system in Eq. (4.47) observable within the vicinity of $\boldsymbol{x}^*$ scaled by $\boldsymbol{U}(\boldsymbol{x}^*, \sigma) = \{x \in \boldsymbol{R}^n \| \boldsymbol{x} - \boldsymbol{x}^*| \le \sigma\}$. The observability in this case is also called local observability.

According to the above lemma, the observability analysis for the nonlinear system can be implemented by investigating the counterpart linearized system.

Let the observability matrix be

$$\boldsymbol{O} = \begin{bmatrix} \boldsymbol{C} \\ \boldsymbol{CA} \\ \vdots \\ \boldsymbol{CA}^{n-1} \end{bmatrix}_{mn \times n}. \tag{4.51}$$

The linearized system is observable, if and only if the column $\boldsymbol{O}$ of observability matrix is of full rank, i.e. the rank of $\boldsymbol{O}$ is n.

Since the practical navigation systems for spacecraft usually employ the measurement provided by navigation sensors to reduce the orbit information predicted by orbital dynamics model of spacecraft, the local observability is sufficient to ensure the success of navigation system. Thus, we will use the linearized observability criterion to investigate the observability of the navigation system with augmented state.

4.4.2.2 Observability Demonstration

We analyze the observability of navigation systems using three X-ray pulsars, and the navigation systems using four pulsars or more can be analyzed in the same way.

Considering the complexity of state model has a little impact on the result of observability analysis, the observability analysis here is performed for a spacecraft suffering only the central-force of a celestial body. In this case, the linearized time-invariant system shown in Eq. (4.52) is investigated.

$$\begin{cases} \delta \dot{\boldsymbol{x}} = \boldsymbol{F} \delta \boldsymbol{x} \\ \delta y = \boldsymbol{H} \delta \boldsymbol{x} \end{cases} \tag{4.52}$$

where

$$\boldsymbol{F} = \begin{bmatrix} \boldsymbol{0}_{3\times3} & \boldsymbol{I}_{3\times3} & \boldsymbol{0}_{3\times3} \\ \boldsymbol{S} & \boldsymbol{0}_{3\times3} & \boldsymbol{0}_{3\times3} \\ \boldsymbol{0}_{3\times3} & \boldsymbol{0}_{3\times3} & \boldsymbol{0}_{3\times3} \end{bmatrix}. \tag{4.53}$$

In Eq. (4.53),

$$\begin{aligned} \boldsymbol{S} &= \left. \frac{\partial \boldsymbol{a}_{TB}}{\partial \boldsymbol{r}} \right|_{t=t_k} \\ &= \begin{bmatrix} \mu_E(3x^2 - \|\boldsymbol{r}\|^2)/\|\boldsymbol{r}\|^5 & 3\mu_E xy/\|\boldsymbol{r}\|^5 & 3\mu_E xz/\|\boldsymbol{r}\|^5 \\ 3\mu_E xy/\|\boldsymbol{r}\|^5 & \mu_E(3y^2 - \|\boldsymbol{r}\|^2)/\|\boldsymbol{r}\|^5 & 3\mu_E yz/\|\boldsymbol{r}\|^5 \\ 3\mu_E xz/\|\boldsymbol{r}\|^5 & 3\mu_E yz/\|\boldsymbol{r}\|^5 & \mu_E(3z^2 - \|\boldsymbol{r}\|^2)/\|\boldsymbol{r}\|^5 \end{bmatrix} \end{aligned} \tag{4.54}$$

In Eq. (4.52),

$$H = \left.\frac{\partial h(X)}{\partial X}\right|_{X=X(t_k)} \approx \begin{bmatrix} n_1 & \mathbf{0}_{1\times 3} & 1\ 0\ 0 \\ n_2 & \mathbf{0}_{1\times 3} & 0\ 1\ 0 \\ n_3 & \mathbf{0}_{1\times 3} & 0\ 0\ 1 \end{bmatrix}. \tag{4.55}$$

And then, the observability matrix is

$$O = \begin{bmatrix} H \\ HF \\ HF^2 \end{bmatrix} = \begin{bmatrix} n_1 & \mathbf{0}_{1\times 3} & 1\ 0\ 0 \\ n_2 & \mathbf{0}_{1\times 3} & 0\ 1\ 0 \\ n_3 & \mathbf{0}_{1\times 3} & 0\ 0\ 1 \\ \mathbf{0}_{1\times 3} & n_1 & 0\ 0\ 0 \\ \mathbf{0}_{1\times 3} & n_2 & 0\ 0\ 0 \\ \mathbf{0}_{1\times 3} & n_3 & 0\ 0\ 0 \\ n_1 S & \mathbf{0}_{1\times 3} & 0\ 0\ 0 \\ n_2 S & \mathbf{0}_{1\times 3} & 0\ 0\ 0 \\ n_3 S & \mathbf{0}_{1\times 3} & 0\ 0\ 0 \end{bmatrix}. \tag{4.56}$$

The rank of O can be calculated as

$$rank(O) = rank(M_1) + rank(M_2) + rank(M_3) \tag{4.57}$$

where

$$M_1 = \begin{bmatrix} 1\ 0\ 0 \\ 0\ 1\ 0 \\ 0\ 0\ 1 \end{bmatrix} \tag{4.58}$$

$$M_2 = \begin{bmatrix} n_1 \\ n_2 \\ n_3 \end{bmatrix} \quad M_3 = \begin{bmatrix} n_1 S \\ n_2 S \\ n_3 S \end{bmatrix}. \tag{4.59}$$

It can be seen from Eq. (4.58) that $rank(M_1)$ equals 3. Moreover, if the three investigated pulsars are not on a plane, n_1, n_2 and n_3 are linearly independent. Therefore, $rank(M_2)$ equals 3, too. Furthermore, the expression of M_3 becomes

$$M_3 = M_2 S. \tag{4.60}$$

Then $|M_3| = |M_2||S|$. From Eq. (4.54), we have $|S| = 2/\|r\|^9$. Thus, $rank(M_3) = 3$.

Finally,

$$rank(O) = rank(M_1) + rank(M_2) + rank(M_3) = 9. \tag{4.61}$$

From Eq. (4.61), O is of full column rank, and the improved navigation system is observable and the augmented state method is feasible.

4.4.3 Simulation Analysis

The navigation pulsars used are PSR B0531+21, B1509−58 and B1821−24. The observation period is 30 min, and the navigation process lasts for 7 days. Assuming that the investigated satellite is on the geosynchronous orbit, the initial position and velocity error are 1 km and 1 m/s respectively. The pulsar angular position error is 1 mas, the ephemeris error of the Earth is 1 km, the spacecraft-borne atomic clock error is 1 μs and the pulsar distance error is 30%.

Figure 4.14 compares the positioning performance of navigation systems with and without error compensation. The positioning estimation error for the system with error compensation could converge into the 3σ profile as the duration of navigation process grows, indicating that the result of the system with the error compensation is unbiased. For the system without error compensation, although the positioning estimation error could reach a steady state, it is out of the 3σ profile. It means that those systematic biases would degrade the positioning result. Thus, the proposed error compensation method can effectively reduce the impact of systematic bias.

Table 4.5 gives the positioning accuracy comparison obtained from 300 Monte-Carlo simulations. The augmented state method could effectively reduce the impact of systematic biases, reaching a positioning accuracy of about 100 m.

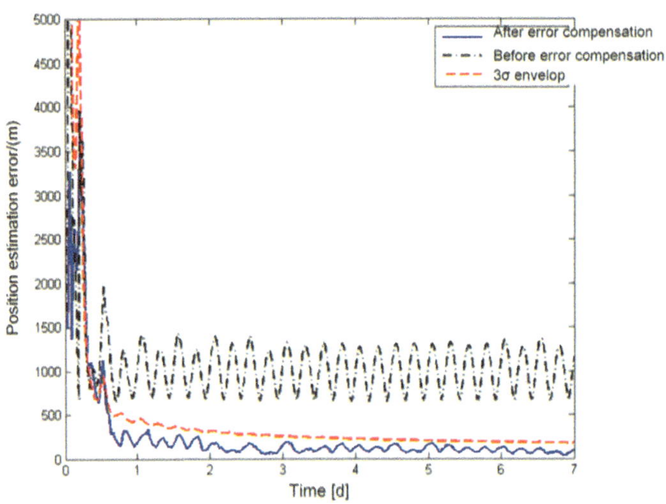

Fig. 4.14 Positioning performance for navigation system with and without error compensation

Table 4.5 Positioning accuracy for systems with and without error compensation

	Positioning accuracy [m]
System with error compensation	96.72
System without error compensation	845.1

4.5 Systematic Biases Compensation Method Based on Time-Differenced Measurement

For augmented state method, the systematic biases are augmented into the state, resulting in that computation for high-dimensional matrices are required, and might cause the computation instability. Regarding that the systematic biases vary slowly over time, this subsection will propose an X-ray pulsar-based navigation method using the time-differenced measurement. This method employs the difference between neighbor epochs as the measurement, and is without the computation of high-dimensional matrices.

4.5.1 *Time-Differenced Measurement Model*

Following from Eq. (4.44), the time-differenced measurement model at t_k can be modeled as

$$\Delta \boldsymbol{Z}_k = \boldsymbol{Z}_k - \boldsymbol{Z}_{k-1} = \boldsymbol{h}(\boldsymbol{x}_k) - \boldsymbol{h}(\boldsymbol{x}_{k-1}). \tag{4.62}$$

The detailed expression of time-differenced measurement model can be obtained by substituting Eq. (4.44) into Eq. (4.62).

Assume that the pulsar angular position error is 1 mas, and the pulsar distant error is 30%. The Earth ephemeris error is approximated by the difference between DE421 and DE405. Figure 4.15 compares the systematic biases within Eq. (4.44) and those within Eq. (4.62) over 2012. Among them, the solid lines stand for systematic biases in Eq. (4.62) and the dotted lines stand for systematic biases in Eq. (4.44). The systematic biases in Eq. (4.62) are far less than those in Eq. (4.44). Therefore, the time-differenced measurement models can greatly reduce the impact of systematic biases.

4.5.2 *Observability Analysis*

In order to ensure the feasibility of time-differenced method, this subsection will perform the observability analysis.

Taking only the central force from the celestial body into consideration, the linearized system with the form of Eq. (4.63) is investigated for the observability analysis.

$$\begin{cases} \dot{\boldsymbol{x}} = \boldsymbol{F}\boldsymbol{x} \\ \boldsymbol{Z} = \boldsymbol{H}_c(\boldsymbol{I}_{6\times 6} - \boldsymbol{\Phi}(t_k, t_{k+1}))\boldsymbol{x} \end{cases} \tag{4.63}$$

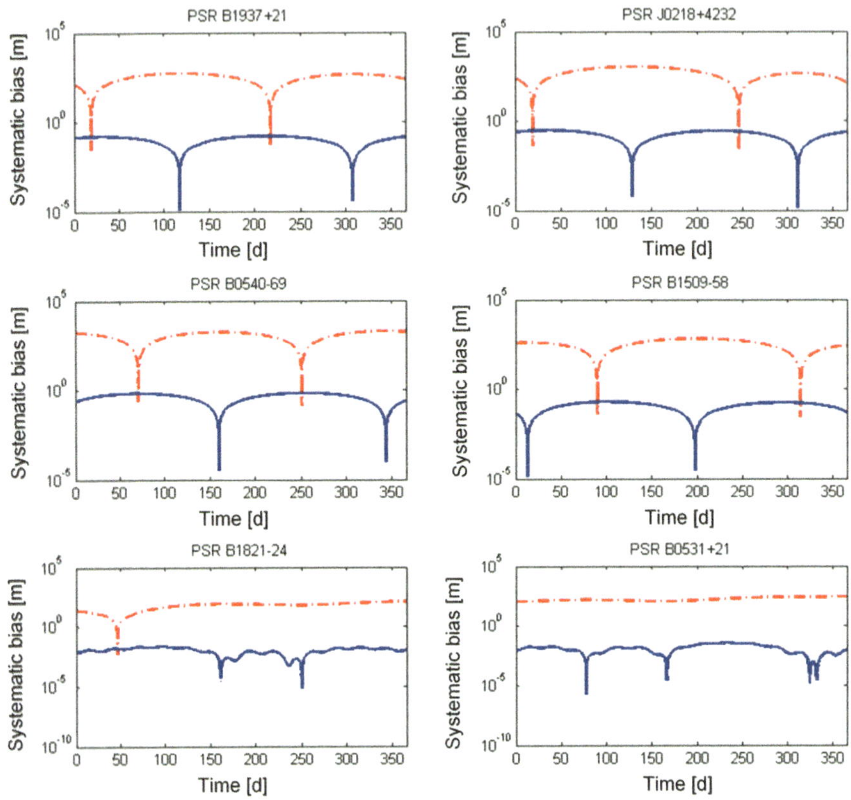

Fig. 4.15 Systematic bias comparison between time-differenced observation models and traditional observation models

where

$$\boldsymbol{F} = \left.\frac{\partial \boldsymbol{f}(\boldsymbol{x})}{\partial \boldsymbol{x}}\right|_{t=t_k} = \begin{bmatrix} \boldsymbol{0}_{3\times3} & \boldsymbol{I}_{3\times3} \\ \boldsymbol{S} & \boldsymbol{0}_{3\times3} \end{bmatrix}. \tag{4.64}$$

In Eq. (4.64), the expression of $\boldsymbol{S}$ is given by Eq. (4.54).

In Eq. (4.63), $\boldsymbol{I}_{6\times6}$ is the unit matrix, and $\boldsymbol{H}_c$ can be expressed as

$$\boldsymbol{H}_c = \begin{bmatrix} \boldsymbol{N}_1 & \boldsymbol{0}_{1\times3} \\ \boldsymbol{N}_2 & \boldsymbol{0}_{1\times3} \\ \boldsymbol{N}_3 & \boldsymbol{0}_{1\times3} \end{bmatrix} \tag{4.65}$$

where $\boldsymbol{N}_j$ ($j = 1, 2, 3$) is the observation matrix of jth observed pulsar, expressed as

$$N_j \approx n_j. \tag{4.66}$$

In Eq. (4.63), $\boldsymbol{\Phi}(t_k, t_{k+1})$ is the state transition matrix, which can be approximated by

$$\boldsymbol{\Phi}(t_k, t_{k+1}) \approx \boldsymbol{I}_{6\times 6} + \boldsymbol{F}(t_k - t_{k+1}). \tag{4.67}$$

Substituting Eq. (4.67) into Eq. (4.63), we have

$$\begin{cases} \dot{\boldsymbol{x}} = \boldsymbol{F}\boldsymbol{x} \\ \boldsymbol{Z} = -(t_k - t_{k+1})\boldsymbol{H}_c\boldsymbol{F}\boldsymbol{x} \end{cases}. \tag{4.68}$$

The observability matrix of the system in Eq. (4.68) is

$$\boldsymbol{O} = \begin{bmatrix} -(t_k - t_{k+1})\boldsymbol{H}_c\boldsymbol{F} \\ -(t_k - t_{k+1})\boldsymbol{H}_c\boldsymbol{F}^2 \end{bmatrix} = (t_{k+1} - t_k)\begin{bmatrix} \boldsymbol{H}_c \\ \boldsymbol{H}_c\boldsymbol{F} \end{bmatrix}\boldsymbol{F} \tag{4.69}$$

Now the determinant of matrix $\boldsymbol{O}$ is

$$|\boldsymbol{O}| = (t_{k+1} - t_k)\left|\begin{bmatrix} \boldsymbol{H}_c \\ \boldsymbol{H}_c\boldsymbol{F} \end{bmatrix}\right||\boldsymbol{F}|. \tag{4.70}$$

In Eq. (4.70), the matrix $\begin{bmatrix} \boldsymbol{H}_c^T & (\boldsymbol{H}_c\boldsymbol{F})^T \end{bmatrix}^T$ is the observability matrix of traditional X-ray pulsar-based navigation system, whose determinant is proved to be not zero. As can be seen from Eq., the determinant of matrix $\boldsymbol{F}$ is not zero either. And then, the determinant of matrix $\boldsymbol{O}$ is not zero, and $\boldsymbol{O}$ is of full column rank. Thus, the X-ray pulsar navigation system using time-differenced measurement model is completely observable.

4.5.3 Modified Unscented Kalman Filter

In the above observability analysis, the process noise and measurement noise are not considered and the state transition matrix is used to model the time-differenced observation model. However, the time-differenced measurement using the orbital dynamics model to connect the measurements at neighbor epochs will make the process and measurement noise become correlated. In addition, the state transition matrix requires computing the Jacobian matrix which is only an approximation to the original system to the first order. Thus, we will develop the Modified Unscented Kalman Filter (MUKF) to overcome the above problems.

Consider the nonlinear system of

$$\boldsymbol{x}_k = \boldsymbol{f}(\boldsymbol{x}_{k-1}) + \boldsymbol{w}_k \tag{4.71}$$

$$\Delta \boldsymbol{Z}_k = \boldsymbol{h}(\boldsymbol{x}_k) - \boldsymbol{h}(\boldsymbol{x}_{k-1}) + \boldsymbol{\upsilon}_k - \boldsymbol{\upsilon}_{k-1} \tag{4.72}$$

where

$$E(\boldsymbol{w}_k) = 0 \ E(\boldsymbol{w}_k \boldsymbol{w}_k^T) = \boldsymbol{Q}_k , \tag{4.73}$$

$$E(\boldsymbol{V}_k) = 0 \ E(\boldsymbol{\upsilon}_k \boldsymbol{\upsilon}_k^T) = \boldsymbol{R}_k \ E(\boldsymbol{w}_k \boldsymbol{\upsilon}_k^T) = \boldsymbol{0} . \tag{4.74}$$

Substituting Eq. (4.71) into Eq. (4.72) yields

$$\begin{aligned} \Delta \boldsymbol{Z}_k &= \boldsymbol{h}(\boldsymbol{f}(\boldsymbol{x}_{k-1}) + \boldsymbol{w}_k) - \boldsymbol{h}(\boldsymbol{x}_{k-1}) + \boldsymbol{\upsilon}_k - \boldsymbol{\upsilon}_{k-1} \\ &= \boldsymbol{G}(\boldsymbol{x}_{k-1}) + \boldsymbol{\upsilon}_k^* \end{aligned} \tag{4.75}$$

where

$$\begin{aligned} \boldsymbol{V}_k^* &= \boldsymbol{h}(\boldsymbol{w}_k) + \boldsymbol{\upsilon}_k - \boldsymbol{\upsilon}_{k-1} \\ &\approx \boldsymbol{H}_k \boldsymbol{w}_k + \boldsymbol{\upsilon}_k - \boldsymbol{\upsilon}_{k-1} , \end{aligned} \tag{4.76}$$

where.

$$\boldsymbol{H}_k = \left. \frac{\partial \boldsymbol{h}(\boldsymbol{x})}{\partial \boldsymbol{x}} \right|_{x=x_k^-} . \tag{4.77}$$

The properties of $\boldsymbol{\upsilon}_k^*$ is

$$\boldsymbol{H}_k = \left. \frac{\partial \boldsymbol{h}(\boldsymbol{x})}{\partial \boldsymbol{x}} \right|_{x=x_k^-} \tag{4.78}$$

$$E(\boldsymbol{\upsilon}_k^*) = 0 \ E(\boldsymbol{\upsilon}_k^* \boldsymbol{\upsilon}_k^{*T}) = \boldsymbol{H}_k \boldsymbol{Q}_k \boldsymbol{H}_k^T + \boldsymbol{R}_k + \boldsymbol{R}_{k-1} \tag{4.79}$$

$$E(\boldsymbol{w}_k \boldsymbol{\upsilon}_k^{*T}) = \boldsymbol{Q}_k \boldsymbol{H}_k^T . \tag{4.80}$$

It can be seen from Eqs. (4.79) and (4.80) that the measurement covariance matrix based on time-differenced measurement model is much larger than the traditional one and its process and measurement noises are correlated.

Assume that the estimated satellite state and the covariance at the $t_{k\text{-}1}$ are

$$\hat{\boldsymbol{x}}_{k-1} = E(\boldsymbol{x}_{k-1}) \ \boldsymbol{P}_{k-1} = E\left[(\boldsymbol{x}_{k-1} - \hat{\boldsymbol{x}}_{k-1})(\boldsymbol{x}_{k-1} - \hat{\boldsymbol{x}}_{k-1})^T\right] \tag{4.81}$$

then the MUKF proceeds as follows

(1) **Time update**

$$\chi_{i,k|k-1} = f(\chi_{i,k-1})\ x_k^- = \sum_{i=0}^{2n} \omega_i^m \chi_{i,k|k-1} \tag{4.82}$$

$$P_k^- = \sum_{i=0}^{2n} \omega_i^c \left[\chi_{i,k|k-1} - x_k^-\right]\left[\chi_{i,k|k-1} - x_k^-\right]^T + Q_k \tag{4.83}$$

where x_k^- and P_k^- are the predicted state and the error covariance matrix at time t_k.

In Eqs. (4.82) and (4.83), the expression of sigma point set $\{\chi_{i,k-1} | i = 0, \ldots, 2n, k \geq 1\}$ is

$$\begin{cases} \chi_{0,k-1} = \hat{x}_{k-1} \\ \chi_{i,k-1} = \hat{x}_{k-1} + \sqrt{n+\xi} \cdot \left(\sqrt{P_{k-1}}\right)_i\ i = 1, 2, \ldots, n \\ \chi_{i+n,k-1} = \hat{x}_{k-1} - \sqrt{n+\xi} \cdot \left(\sqrt{P_{k-1}}\right)_i\ i = n+1, n+2, \ldots, 2n \end{cases} \tag{4.84}$$

where n is the dimension of state vector, and $\xi = \alpha^2(n + \kappa) - n$ where α is used to control the distribution of sigma points whose value ranges from 0 to 1, and $\kappa = 3 - n$, $\sqrt{P_{k-1}}$ is the Cholesky factor of P_{k-1}.

The weights of mean value and variance are given by

$$\begin{cases} \omega_0^m = \xi/(n+\xi) \\ \omega_0^c = \xi/(n+\xi) + \left(1 - \alpha^2 + \beta\right) \\ \omega_i^m = \omega_i^c = \xi/[2(n+\xi)] \end{cases} \tag{4.85}$$

where β is the parameter related to the prior distribution of state. For Gaussian distribution, the optimal value of β is 2.

(2) **Measurement update**

$$\Delta Z_{i,k|k-1} = h(\chi_{i,k|k-1}) - h(\chi_{i,k-1})\ \Delta Z_k^- = \sum_{i=0}^{2n} \omega_i^m \Delta Z_{i,k|k-1} \tag{4.86}$$

$$\begin{aligned} P_{\tilde{Z}_k \tilde{Z}_k} &= \sum_{i=0}^{2n} \omega_i^c \left[\Delta Z_{i,k|k-1} - \Delta Z_k^-\right]\left[\Delta Z_{i,k|k-1} - \Delta Z_k^-\right]^T + E(V_k^* V_k^{*T}) \\ &= \sum_{i=0}^{2n} \omega_i^c \left[\Delta Z_{i,k|k-1} - \Delta Z_k^-\right]\left[\Delta Z_{i,k|k-1} - \Delta Z_k^-\right]^T \\ &\quad + H_k Q_k H_k^T + R_k + R_{k-1} \end{aligned} \tag{4.87}$$

$$
\begin{aligned}
P_{\tilde{x}_k\tilde{Z}_k} &= \sum_{i=0}^{2n} \omega_i^c \left[\chi_{i,k|k-1} - x_k^-\right]\left[\Delta Z_{i,k|k-1} - \Delta Z_k^-\right]^T + E(w_k V_k^{*T}) \\
&= \sum_{i=0}^{2n} \omega_i^c \left[\chi_{i,k|k-1} - x_k^-\right]\left[\Delta Z_{i,k|k-1} - \Delta Z_k^-\right]^T + Q_k H_k^T
\end{aligned} \quad (4.88)
$$

$$
\hat{x}_k = x_k^- + K_k\left(\Delta Z_k - \Delta Z_k^-\right)\ \hat{P}_k = P_k^- - K_k P_{\tilde{Z}_k\tilde{Z}_k} K_k^T \quad (4.89)
$$

$$
K_k = P_{\tilde{x}_k\tilde{Z}_k} P_{\tilde{Z}_k\tilde{Z}_k}^{-1} \quad (4.90)
$$

where $\hat{x}_k$ and $\hat{P}_k$ are the predicted state and the error covariance matrix at the time t_k.

As can be seen from Eqs. (4.83)–(4.90), for the MUKF, the time-differenced measurement model is established by propagating sigma points, and the linearization error can be reduced and the impact of the correlation between process noise and measurement noise is well tackled.

4.5.4 Simulation Analysis

In this section, three satellites, OPS_5111, GGTS and INTELSAT_2-F2, are used for analyzing the feasibility of the time-differenced algorithm. The initial orbital elements of these three satellites are listed in Table 4.6.

PSR B1937+21, PSR B1821−24 and PSR B0531+21 are selected as the navigation pulsars. Assume that the pulsar angular position error is 1 mas and the pulsar distance error is 30%. The standard covariance of measurement noise is determined by Eq. (5.3), where the sensor area is 1 m^2 and the observation period of pulsar is 1800 s.

The DE405 ephemeris is used to provide the Earth position over the whole navigation process, and the difference between DE421 and DE405 is used to approximate

Table 4.6 Initial orbital elements of three satellites

Orbital element	Satellite name		
	OPS_5111	GGTS	INTELSAT_2-F2
Semi-major axis [km]	39,781.66	40,133.97	42,199.55
Eccentricity	0.0051	0.0027	0.0015
Inclination [°]	10.01	4.26	10.44
Right ascension of ascending node [°]	330.38	292.19	322.41
Argument of perigee angular distance [°]	195.92	199.77	201.63
Mean anomaly [°]	10.15	356.26	184.45

the error in DE405. The initial clock error of satellite atomic clock is 1 μs, and the clock frequency drift rate is 10^{-11} Hz. The initial position and velocity errors are 1 km and 2 m/s respectively.

Figure 4.16 shows the positioning performance comparison between the time-differenced method and the method without error compensation. Both methods ensure filtering convergence, but the time-differenced method can achieve a higher positioning accuracy.

Table 4.7 gives the results of 300 Monte-Carlo simulations. Table 4.7 shows that the time-differenced method can effectively reduce the impact of systematic biases. However, the positioning results for satellite INTELSAT_2-F2 is the worst among the three satellites. In order to adequately demonstrate the effectiveness of time-differenced method, the analysis in the remainder part focuses on the performance on satellite INTELSAT_2-F2.

Figure 4.17 compares the performances of MUKF, UKF and EKF. Figure 4.17 shows that UKF, MUKF and EKF all can converge. However, compared to EKF, the final states of MUKF and UKF are steadier. This is because both MUKF and

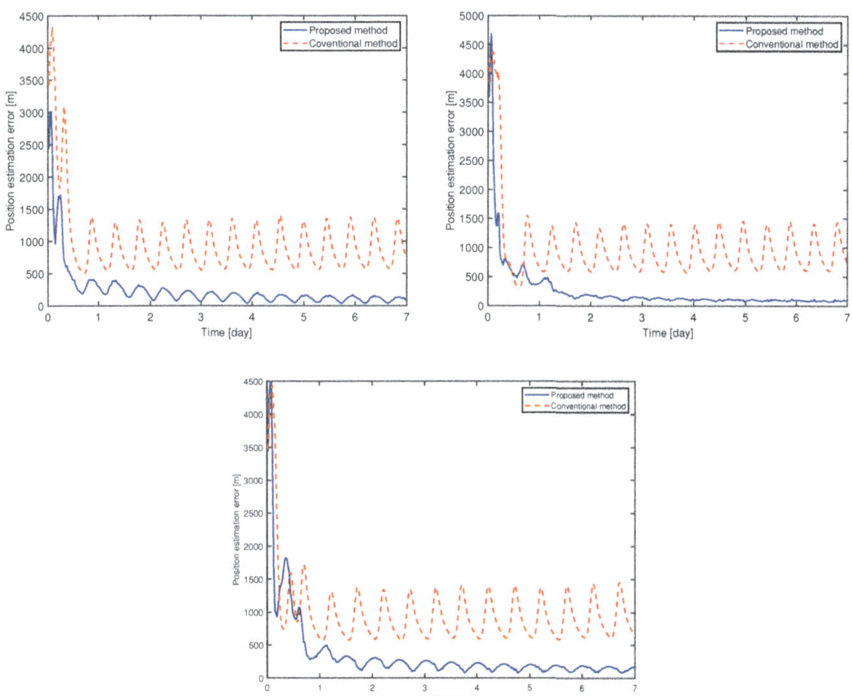

Fig. 4.16 Performance comparison between time-differenced method and the method without systematic bias compensation

Table 4.7 Mean of position estimation accuracy [m]

Navigation method	Satellite name		
	OPS_5111	GGTS	INTELSAT_2-F2
Time-differenced method	107.64	91.36	137.98
Method without systematic bias compensation	904.06	898.23	921.36

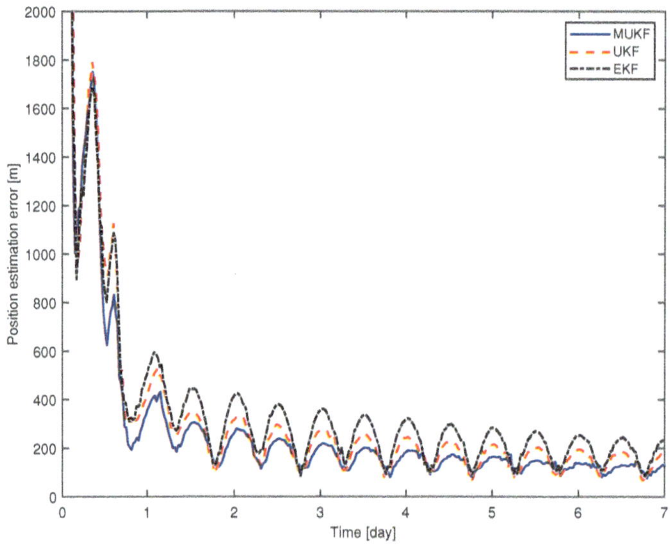

Fig. 4.17 Performance comparison among MUKF, UKF and EKF

UKF employ the selected sigma points to generate the time-differenced observation model without involving linearization error, while the linear error in EKF will affect its performance. Additionally, MUKF has a performance better than UKF, because MUKF considers the correlation between process and measurement noise but UKF does not.

Figure 4.18 compares the positioning performance among the time-differenced method, the augmented state method and the method proposed in Ref. [5], which is refered as "method with smoother". All methods can reduce the impact of systematic bias. However, the final positioning accuracies of the time-differenced method and the augmented state method are superior to that of the method proposed in Ref. [5]. Compared to the augmented state method, the method described in this section has a faster convergence speed and saves on computation.

When only the pulsar angular position error is considered, Fig. 4.19 gives the positioning results varies with different pulsar angular position errors. The positioning estimation error grows as the angular position error increases. However, for the

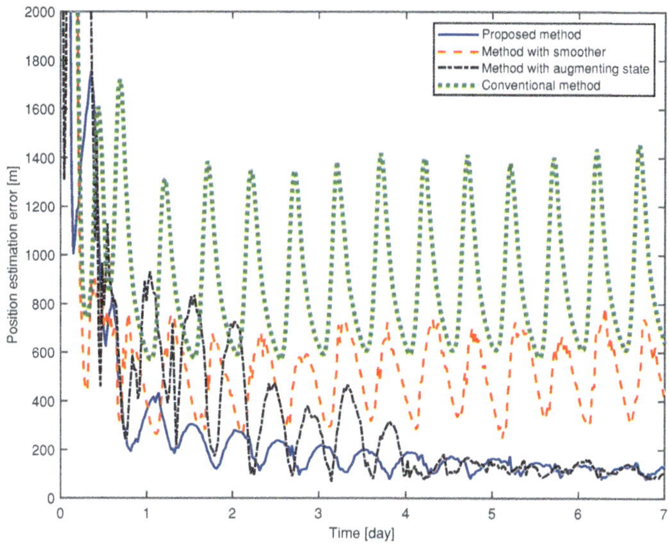

Fig. 4.18 Comparison among three error compensation methods

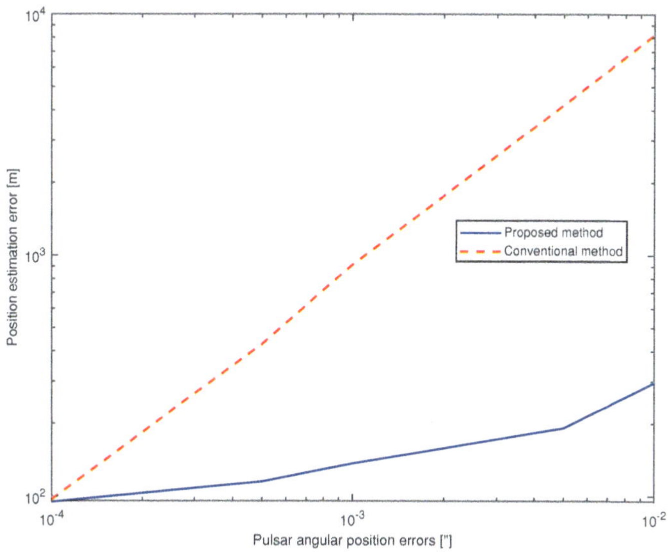

Fig. 4.19 Variance of position estimation error with pulsar angular position error

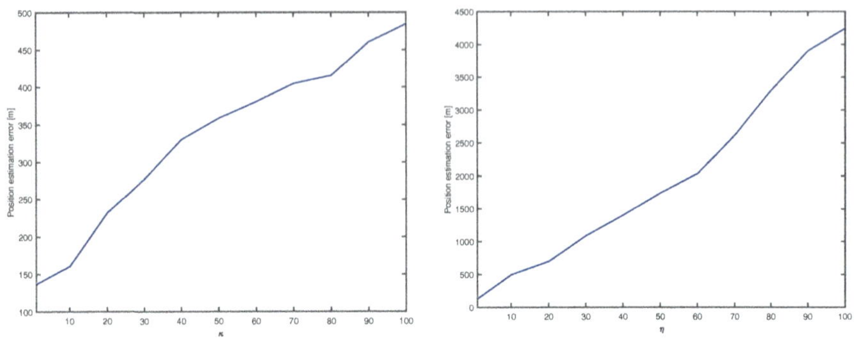

Fig. 4.20 Relation between Position estimation error and process noise and measurement noise

same angular position error, the methods proposed in this section are superior to the traditional one. Therefore, the time-differenced method could resist the impact of pulsar angular position error to some extent. The conclusions for the other error sources could be obtained in a similar way.

As Eq. (4.79) shows, the time-differenced measurement model would enlarge the measurement noise. In addition, the process noise also affects the positioning performance of the proposed method. It is necessary to analyze the impact of the process and measurement noise on the positioning performance of time-differenced method.

Considering that process and measurement noise are modeled as zero-mean Gaussian white noises. We could produce different process noises by multiplying the standard covariance of process noise by a coefficient κ and different measurement noises by multiplying the standard covariance of measurement noise by coefficient η. Figure 4.20 shows the positioning estimation error with different coefficients. The positioning estimation error grows as the process noise and measurement noise increase. Besides, the time-differenced method is more sensitive to the growth of measurement noise. If η is more than 200, the positioning estimation error will be more than 1000 m, and the compensation of the error is not obvious in such case. Therefore, the navigation pulsars for time-differenced method should be selected carefully.

4.6 Summary

This chapter, starting from the fundamental principle of X-ray pulsar-based navigation, respectively models the error sources influencing the TOA measurement accuracy and those influencing the accuracy of the time transfer model. Then it analyzes how the TOA measurement accuracy influences the final navigation accuracy and time-keeping accuracy through covariance analysis. According to the results, if systematic biases are neglected and the TOA measurement accuracy is higher than

1 μs, the final positioning accuracy can be about 100 m and the time-keeping accuracy can be about 0.1 μs. This chapter, through Monte Carlo simulations, analyzes how systematic biases impact the accuracy of the time transfer model and the final positioning and time-keeping accuracies. Results show that, for LEO spacecraft, the Earth ephemeris position error, pulsar's angular position error, pulsar distance error and spacecraft-borne atomic clock error are main error sources and can cause the deviation of the positioning system to be hundreds of meters. Hence, to essentially enhance the performance of the X-ray pulsar-based navigation system, it is important to compensate those four error sources.

References

1. Folkner WM, William JG, Boggs DH (2009) The planetary and lunar ephemeris DE 421. IPN Progress Report 42-178, 2009
2. Liu J, Ma J, Tian J et al (2010) X-ray pulsar navigation method for spacecraft with pulsar direction error. Adv Space Res 46:1409–1417
3. http://www.atnf.csiro.au/research/pulsar/psrcat/ (2010)
4. Tian F (2011) Error estimation of pulsar timing based spacecraft autonomous positioning. Prog Astron 29(1):97–104
5. Liu J, Ma J, Tian W (2010) CNS/pulsar integrated navigation using two-level filter. Chin J Electron 19(2):265–269

Chapter 5
X-Ray Pulsar/Multiple Measurement Information Fused Navigation

As the pulsar signal is extremely weak and discontinuous, the spacecraft needs to observe the target pulsar for a long duration to accumulate photons enough to extract the pulse TOA [1]. During the observation period of pulsar, the position of spacecraft has to be predicted by propagating the orbital dynamics model, resulting in the position error presents a nonlinear growth. Additionally, the low-thrust deep space explorers cannot autonomously position by only fusing the information of dynamics and pulsar measurement.

Regarding that different navigation methods have distinctive properties, it is feasible to combine the X-ray pulsar measurement with the other types of measurement information to solve the above problems. According to whether the information fusion needs to change the dynamical model, this chapter proposes two integrated navigation frameworks: (1) X-ray pulsar-based navigation (XNAV)/traditional celestial navigation system (CNS) integrated navigation framework; and (2) XNAV/ INS integrated navigation framework.

5.1 XNAV/CNS Integrated Navigation Framework

CNS measurements are direction and position of the spacecraft with respect to the celestial body, and CNS has the advantage of short signal sampling period [2]. Therefore, the integration of traditional celestial observation information into XNAV might reduce the accumulated propagation error during the observation period of pulsar and accelerate the convergence of filter.

According to the properties of Earth-orbiting spacecraft and deep space explorer, we propose two integrated navigation methods, including XNAV/stellar angle integrated navigation method and XNAV/Sun information integrated navigation method.

W. Zheng and Y. Wang, *X-ray Pulsar-based Navigation*, Navigation: Science and Technology 5, https://doi.org/10.1007/978-981-15-3293-1_5

5.1.1 Traditional Celestial Measurement Model

5.1.1.1 Stellar Angle Measurement Model

The so-called stellar angle measurement refers to the angle between the direction vector of a reference planet and of a star with respect to the satellite. For Earth-orbiting spacecraft, the Earth can be selected as the reference planet. In this case, the diagram of stellar angle is shown in Fig. 5.1. In the figure, α is the stellar angle measurement, $\boldsymbol{s}$ is the direction vector of the reference star which can be obtained by a star sensor, and $\boldsymbol{r}_{S/E}$ is the position vector of spacecraft with respect to the geocenter whose direction vector can be obtained by a horizon sensor.

Then, the stellar angle measurement model is given by [2]

$$\alpha = Z_{st} = \boldsymbol{h}_{st}(\boldsymbol{x}) + v_{\alpha} = \arccos\left(-\frac{\boldsymbol{r}_{S/E} \cdot \boldsymbol{s}}{\left\|\boldsymbol{r}_{S/E}\right\|}\right) + \upsilon_{\alpha} \tag{5.1}$$

where υ_{α} is the measurement noise, which can be modeled as a zero-mean Gaussian white noise whose variance is determined by accuracies of star sensor and horizon sensor.

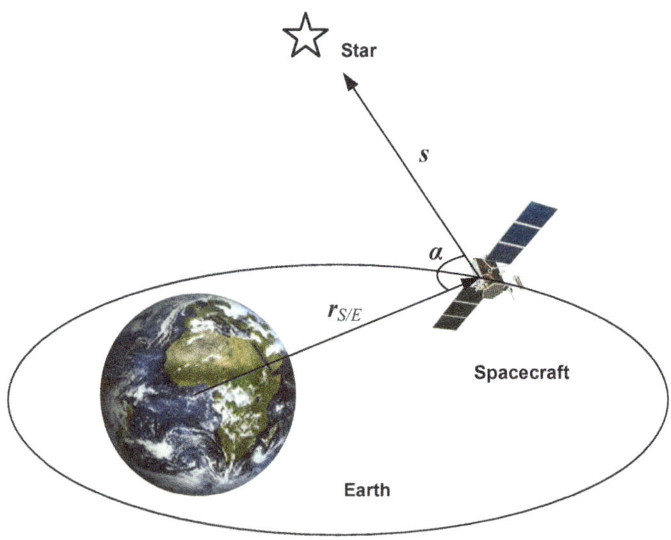

Fig. 5.1 Schematic diagram of stellar angle position measurement

5.1.1.2 Sun Information Measurement Model

The Sun information measurement model includes the Sun direction vector and the radial Doppler velocity [3]. The Sun direction vector can be measured by a Sun sensor, and the radial Doppler velocity can be obtained by a spectrometer. The spectrometer obtains the Doppler shift by sunlight imaging and then measurement of spectral line shift generated by the relative motion of Sun and the spacecraft. The Sun information measurement principle is shown in Fig. 5.2. In the figure, $\boldsymbol{r}_{S/S}$ is the position vector of the deep space explorer with respect to the barycenter of Sun.

According to Fig. 5.2, the measurement models of the Sun direction vector and radial Doppler velocity can be expressed as [4]

$$\boldsymbol{Y}_{LOS} = \boldsymbol{h}_{LOS}(\boldsymbol{x}) + \boldsymbol{\upsilon}_{LOS} = \frac{1}{\left\|\boldsymbol{r}_{S/S}\right\|}\boldsymbol{r}_{S/S} + \boldsymbol{\upsilon}_{LOS} \tag{5.2}$$

$$Y_D = \boldsymbol{h}_D(\boldsymbol{x}) + \upsilon_D = \frac{1}{\left\|\boldsymbol{r}_{S/S}\right\|}\dot{\boldsymbol{r}}_{S/S} \cdot \boldsymbol{r}_{S/S} + \upsilon_D. \tag{5.3}$$

In Eq. (5.2), $\boldsymbol{Y}_{LOS}$ is the Sun direction vector and $\boldsymbol{\upsilon}_{LOS}$ is the measurement noise which can be modeled as the zero-mean Gaussian white noise with a variance determined by the accuracy of Sun sensor. In Eq. (5.3), Y_D is the radial Doppler velocity, and υ_D is the measurement noise which might be modeled as the zero-mean Gaussian white noise with a variance determined by the accuracy of spectrometer.

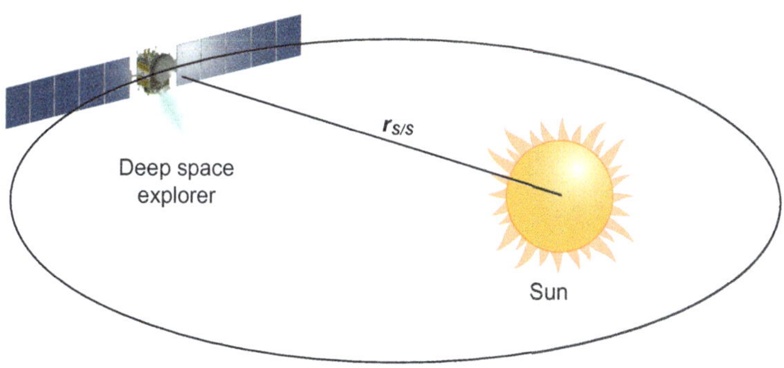

Fig. 5.2 Schematic diagram of Sun information measurement

5.1.2 Information Fusion Method

5.1.2.1 Federated Filter

The federated filter is a two-stage distributed filtering method, which comprises several sub-filters and one main filter. Time updating and measure updating of sub-filters are independent. And the main filter fuses results provided by sub-filters. The result of main filter is fed back to the sub-filter as the initial value of next period of filter [5].

We take two sensors as an example to illustrate the work of federated filter. The typical structure of federated filter is shown in Fig. 5.3.

Besides the structure shown in Fig. 5.3, the feedback from the main filter to sub-filters might be omitted in order to prevent the mutual information interference between sub-filters. However, the federated filter with such structure can hardly obtain ideal information fusion results. Therefore, only the structure of federated filter shown in Fig. 5.3 is analyzed in this chapter.

In Fig. 5.3, $\boldsymbol{y}_{1,k}$ and $\boldsymbol{y}_{2,k}$ are measurements of sensor #1 and sensor #2 at the time t_k, $\boldsymbol{x}^-$ is the predicted state obtained by the dynamical model, $\boldsymbol{x}_{1,k}^+$ and $\boldsymbol{x}_{2,k}^+$ are the posterior states obtained by using the observed data from sensor #1 and sensor #2, and $\boldsymbol{P}_{1,k}$ and $\boldsymbol{P}_{2,k}$ are corresponding state error covariance matrices. By fusing the results of the two sub-filters, the result of the main filter can be expressed as

$$\boldsymbol{x}_{g,k} = [\boldsymbol{P}_{1,k}^{-1} + \boldsymbol{P}_{2,k}^{-1}]^{-1}[\boldsymbol{P}_{1,k}^{-1}\boldsymbol{x}_{1,k}^{+} + \boldsymbol{P}_{2,k}^{-1}\boldsymbol{x}_{2,k}^{+}] \tag{5.4}$$

$$\boldsymbol{P}_{g,k} = [\boldsymbol{P}_{1,k}^{-1} + \boldsymbol{P}_{2,k}^{-1}]^{-1}. \tag{5.5}$$

The main filter feeds back to the sub-filters according to Eqs. (5.6)–(5.8).

$$\boldsymbol{x}_{i,k} = \boldsymbol{x}_{g,k} (i = 1, 2) \tag{5.6}$$

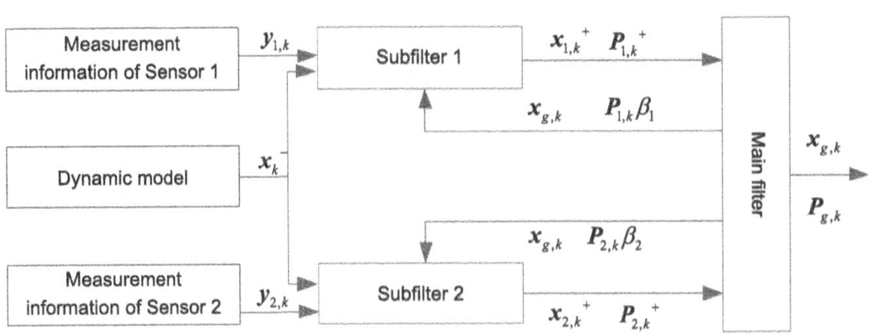

Fig. 5.3 Structure of federated filter

$$P_{i,k} = P_{g,k}\beta_i \tag{5.7}$$

$$\beta_i = \frac{\left\|P_{i,k}\right\|^{-1}}{\left\|P_{1,k}\right\|^{-1} + \left\|P_{2,k}\right\|^{-1}}. \tag{5.8}$$

Meanwhile, $x_{1,k}^{+}$ and $x_{2,k}^{+}$ can be given by

$$x_{i,k}^{+} = K_{i,k}y_{i,k} + (I - K_{i,k}H_{i,k})x_k^{-} (i = 1, 2) \tag{5.9}$$

where K_i is the Kalman gain matrix, H_i is the observation matrix of sub-filter, I is the unit matrix of $n \times n$, and n is the dimension of state vector x.

According to the covariance theorem, we obtain [6]

$$\mathrm{cov}(x_{1,k}^{+}, x_{2,k}^{+}) = (I - K_{1,k}H_{1,k})P_k^{-}(I - K_{2,k}H_{2,k}). \tag{5.10}$$

As can be seen from Eq. (5.10), results of sub-filters are correlated. However, the information fusion of the main filter is based on the least square criterion whose inputs are assumed to be not correlated with each other. Obviously, the current federated filter does not meet such requirement, and it is unable to obtain the globally optimal solution. Therefore, it is necessary to find an information fusion algorithm that might obtain the globally optimal solution.

5.1.2.2 Kinematic and Static Filter

(1) **Basic principle**

To avoid the problem that the output of a federated filter is suboptimal, Yang [7] proposed an information fusion filter named "kinematic and static filter" and proved that the proposed filter is globally optimal. The kinematic and static filter is composed of a kinematic filter and a static filter. The kinematic filter works based on the data from the dynamic model and the measurement from one measurement sensor, and the static filter works based on the output of the kinematic filter and the measurements from the other sensors. Therefore, the data from the dynamic model is just used once and the output of the static filter is globally optimal. For the case that two sensors are used, Fig. 5.4 shows the structure of the kinematic and static filter.

According to Fig. 5.4, the predicted spacecraft state x_k^{-} at time t_k and the corresponding error covariance matrix P_k^{-} are provided by the dynamic model. The measurement from sensor 1 is $y_{1,k}$, whose error covariance matrix is $R_{1,k}$. The solution of the kinematic filter can be derived as

$$x_{1,k}^{+} = [(P_k^{-})^{-1} + H_{1,k}^T(R_{1,k})^{-1}H_{1,k}]^{-1}[(P_k^{-})^{-1}x_k^{-} + H_{1,k}^T(R_{1,k})^{-1}y_{1,k}] \tag{5.11}$$

$$P_{1,k}^{+} = (I - K_{1,k}H_{1,k})P_k^{-} \tag{5.12}$$

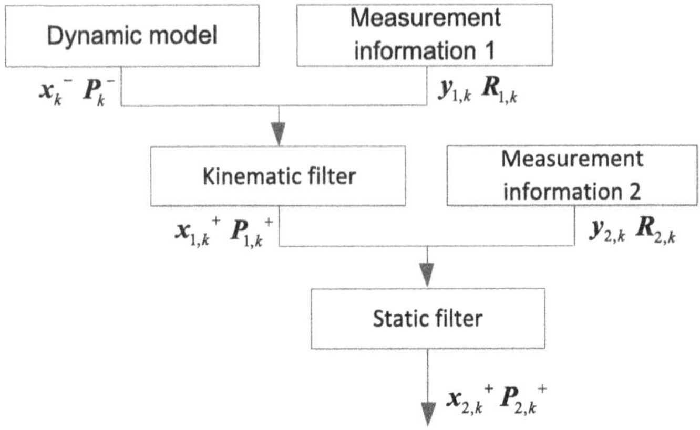

Fig. 5.4 Structure of kinematic and static filter

where

$$\boldsymbol{K}_{1,k} = \boldsymbol{P}_k^- \boldsymbol{H}_{1,k}^T (\boldsymbol{H}_{1,k} \boldsymbol{P}_k^- \boldsymbol{H}_{1,k}^T + \boldsymbol{R}_{1,k})^{-1}. \tag{5.13}$$

In Eqs. (5.11)–(5.13), $\boldsymbol{H}_{1,k}^T$ is the measurement matrix of $\boldsymbol{y}_{1,k}$.

At the stage of static filter, the spacecraft state predicted by the dynamic model is no longer used. Instead the result of the kinematic filter, $\boldsymbol{x}_{1,k}^+$ and $\boldsymbol{P}_{1,k}^+$, are utilized to fuse with the measurement from sensor 2, $\boldsymbol{y}_{2,k}$, whose error covariance matrix is $\boldsymbol{R}_{2,k}$.

The state model of static filter is

$$\boldsymbol{x}_{2,k}^- = \boldsymbol{x}_{1,k}^+ \; \boldsymbol{P}_{2,k}^- = \boldsymbol{P}_{1,k}^+ . \tag{5.14}$$

Based on Eq. (5.14), the result of static filter stage can be presented as

$$\boldsymbol{x}_{2,k}^+ = [(\boldsymbol{P}_{2,k}^-)^{-1} + \boldsymbol{H}_{2,k}^T (\boldsymbol{R}_{2,k})^{-1} \boldsymbol{H}_{2,k}]^{-1} [(\boldsymbol{P}_{2,k}^-)^{-1} \boldsymbol{x}_{2,k}^- + \boldsymbol{H}_{2,k}^T (\boldsymbol{R}_{2,k})^{-1} \boldsymbol{y}_{2,k}] \tag{5.15}$$

$$\boldsymbol{P}_{2,k}^+ = (\boldsymbol{I} - \boldsymbol{K}_{2,k} \boldsymbol{H}_{2,k}) \boldsymbol{P}_{2,k}^- \tag{5.16}$$

where

$$\boldsymbol{K}_{2,k} = \boldsymbol{P}_{2,k}^- \boldsymbol{H}_{2,k}^T (\boldsymbol{H}_{2,k} \boldsymbol{P}_{2,k}^- \boldsymbol{H}_{2,k}^T + \boldsymbol{R}_{2,k})^{-1}. \tag{5.17}$$

(2) **Unscented kinematic and static filter**

The kinematic and static filter was initially proposed for the linear system and was later extended to a non-linear system by Dr. Yang applying the concept of extended Kalman filter (EKF) [8]. However, EKF needs to derive the Jacobian matrix, and the corresponding linearization error is hardly negligible. For this reason, we derive the unscented kinematic and static filter to reduce the impact of the linearization error by introducing the concept of unscented Kalman filter (UKF) into the kinematic and static filter [9].

Suppose that the system model of integrated navigation system is

$$\boldsymbol{x}_k = \boldsymbol{f}(\boldsymbol{x}_{k-1}) + \boldsymbol{w}_k \tag{5.18}$$

$$\begin{bmatrix} \boldsymbol{Z}_{1,k} \\ \boldsymbol{Z}_{2,k} \end{bmatrix} = \begin{bmatrix} \boldsymbol{h}_1(\boldsymbol{x}_k) \\ \boldsymbol{h}_2(\boldsymbol{x}_k) \end{bmatrix} + \begin{bmatrix} \boldsymbol{v}_{1,k} \\ \boldsymbol{v}_{2,k} \end{bmatrix} \tag{5.19}$$

where

$$E(\boldsymbol{w}_k) = 0 \; E(\boldsymbol{w}_k \boldsymbol{w}_k^T) = \boldsymbol{Q}_k \tag{5.20}$$

$$E(\boldsymbol{v}_{1,k}) = 0 \; E(\boldsymbol{v}_{1,k} \boldsymbol{v}_{1,k}^T) = \boldsymbol{R}_{1,k} \tag{5.21}$$

$$E(\boldsymbol{v}_{2,k}) = 0 \; E(\boldsymbol{v}_{2,k} \boldsymbol{v}_{2,k}^T) = \boldsymbol{R}_{2,k} \, . \tag{5.22}$$

Assume mean value and variance of satellite state at time t_{k-1} are

$$\hat{\boldsymbol{x}}_{k-1} = E(\boldsymbol{x}_{k-1}) \; \boldsymbol{P}_{k-1} = E\left[(\boldsymbol{x}_{k-1} - \hat{\boldsymbol{x}}_{k-1})(\boldsymbol{x}_{k-1} - \hat{\boldsymbol{x}}_{k-1})^T\right] \tag{5.23}$$

Then the improved unscented kinematic and static filter proceeds with the following equations

1. **Kinematic filter**

   (a) **Time update**

$$\boldsymbol{\chi}_{i,k|k-1} = \boldsymbol{f}(\boldsymbol{\chi}_{i,k-1}) \; \boldsymbol{x}_k^- = \sum_{i=0}^{2n} \omega_i^m \boldsymbol{\chi}_{i,k|k-1} \tag{5.24}$$

$$\boldsymbol{P}_k^- = \sum_{i=0}^{2n} \omega_i^c \left[\boldsymbol{\chi}_{i,k|k-1} - \boldsymbol{x}_k^-\right]\left[\boldsymbol{\chi}_{i,k|k-1} - \boldsymbol{x}_k^-\right]^T + \boldsymbol{Q}_k \tag{5.25}$$

where $\boldsymbol{x}_k^-$ and $\boldsymbol{P}_k^-$ are the predicted satellite state and corresponding error covariance matrix at time t_k.

In Eqs. (5.24) and (7.24), the expression of sigma point set $\{\boldsymbol{\chi}_{i,k-1} | i = 0, \ldots, 2n, k \geq 1\}$ is

$$\begin{cases} \boldsymbol{\chi}_{0,k-1} = \hat{\boldsymbol{x}}_{k-1} \\ \boldsymbol{\chi}_{i,k-1} = \hat{\boldsymbol{x}}_{k-1} + \sqrt{n+\xi} \cdot \left(\sqrt{\boldsymbol{P}_{k-1}}\right)_i \; i = 1, 2, \ldots, n \\ \boldsymbol{\chi}_{i+n,k-1} = \hat{\boldsymbol{x}}_{k-1} - \sqrt{n+\xi} \cdot \left(\sqrt{\boldsymbol{P}_{k-1}}\right)_i \; i = n+1, n+2, \ldots, 2n \end{cases} \tag{5.26}$$

where n is the dimension of the state vector, $\xi = \alpha^2(n+\kappa) - n$, where α is used for controlling the distribution of sigma points, whose value ranges from 0 to 1; $\kappa = 3 - n$, and $\sqrt{\boldsymbol{P}_{k-1}}$ is the Cholesky factor of $\boldsymbol{P}_{k-1}$.

The weights of mean value and variance are

$$\begin{cases} \omega_0^m = \xi/(n+\xi) \\ \omega_0^c = \xi/(n+\xi) + \left(1 - \alpha^2 + \beta\right) \\ \omega_i^m = \omega_i^c = \xi/[2(n+\xi)] \end{cases} \tag{5.27}$$

where β is the parameter related to the prior distribution of state. For Gaussian distribution, the optimal value of β is 2.

(b) **Measurement update**

$$\boldsymbol{Z}_{i,k|k-1}^1 = \boldsymbol{h}_1(\boldsymbol{\chi}_{i,k|k-1})\ \boldsymbol{Z}_{1,k}^- = \sum_{i=0}^{2n} \omega_i^m \boldsymbol{Z}_{i,k|k-1}^1 \tag{5.28}$$

$$\boldsymbol{P}_{1,\tilde{\boldsymbol{Z}}_{1,k}\tilde{\boldsymbol{Z}}_{1,k}} = \sum_{i=0}^{2n} \omega_i^c \left[\boldsymbol{Z}_{i,k|k-1}^1 - \boldsymbol{Z}_{st,k}^-\right]\left[\boldsymbol{Z}_{i,k|k-1}^1 - \boldsymbol{Z}_{st,k}^-\right]^T + \boldsymbol{R}_{1,k} \tag{5.29}$$

$$\boldsymbol{P}_{1,\tilde{\boldsymbol{x}}_k\tilde{\boldsymbol{Z}}_{1,k}} = \sum_{i=0}^{2n} \omega_i^c \left[\boldsymbol{\chi}_{i,k|k-1} - \boldsymbol{x}_k^-\right]\left[\boldsymbol{Z}_{i,k|k-1}^1 - \boldsymbol{Z}_{1,k}^-\right]^T \tag{5.30}$$

$$\boldsymbol{x}_{1,k}^+ = \boldsymbol{x}_k^- + \boldsymbol{K}_{st,k}\left(\boldsymbol{Z}_{1,k} - \hat{\boldsymbol{Z}}_{1,k}^-\right)\ \boldsymbol{P}_{1,k}^+ = \boldsymbol{P}_k^- - \boldsymbol{K}_{1,k}\boldsymbol{P}_{1,\tilde{\boldsymbol{Z}}_{1,k}\tilde{\boldsymbol{Z}}_{1,k}}\boldsymbol{K}_{1,k}^T \tag{5.31}$$

$$\boldsymbol{K}_{1,k} = \boldsymbol{P}_{1,\tilde{\boldsymbol{x}}_k\tilde{\boldsymbol{Z}}_{1,k}}\boldsymbol{P}_{1,\tilde{\boldsymbol{Z}}_{1,k}\tilde{\boldsymbol{Z}}_{1,k}}^{-1} \tag{5.32}$$

where $\boldsymbol{x}_{1,k}^+$ and $\boldsymbol{P}_{1,k}^+$ are results of the kinematic filter at the time t_k.

2. **Static filter**

Based on Eq. (5.14), we obtain the input of static filter, namely

$$\boldsymbol{x}_{2,k}^- = \boldsymbol{x}_{1,k}^+\ \boldsymbol{P}_{2,k}^- = \boldsymbol{P}_{1,k}^+ \tag{5.33}$$

performs the unscented transformation again by using Eq. (5.33), namely,

$$\begin{cases} \boldsymbol{\varepsilon}_{0,k-1} = \boldsymbol{x}_{p,k}^- \\ \boldsymbol{\varepsilon}_{i,k-1} = \boldsymbol{x}_{p,k}^- + \sqrt{n+\xi}\cdot\left(\sqrt{\boldsymbol{P}_{p,k}^-}\right)_i\ i = 1, 2, \ldots, n \\ \boldsymbol{\varepsilon}_{i+n,k-1} = \boldsymbol{x}_{p,k}^- - \sqrt{n+\xi}\cdot\left(\sqrt{\boldsymbol{P}_{p,k}^-}\right)_i\ i = n+1, n+2, \ldots, 2n \end{cases} \tag{5.34}$$

where $\sqrt{\boldsymbol{P}_{p,k}^-}$ is the Cholesky factor of $\boldsymbol{P}_{p,k-1}^-$.

The static filter only includes the measurement update which can be expressed as

$$Z^2_{i,k|k-1} = h_2(\varepsilon_{i,k|k-1})\; Z^-_{2,k} = \sum_{i=0}^{2n} \omega_i^m Z^2_{i,k|k-1} \tag{5.35}$$

$$P_{2,\tilde{Z}_{p,k}\tilde{Z}_{2,k}} = \sum_{i=0}^{2n} \omega_i^c \left[Z^2_{i,k|k-1} - Z^-_{2,k}\right]\left[Z^2_{i,k|k-1} - Z^-_{2,k}\right]^T + R_{2.k} \tag{5.36}$$

$$P_{2,\tilde{x}_k\tilde{Z}_{2,k}} = \sum_{i=0}^{2n} \omega_i^c \left[\varepsilon_{i,k|k-1} - x^-_{2,k}\right]\left[Z^2_{i,k|k-1} - Z^-_{2,k}\right]^T \tag{5.37}$$

$$x^+_{2,k} = x^-_{2,k} + K_{2,k}\left(Z_{2,k} - Z^-_{2,k}\right)\; P^+_{2,k} = P^-_{2,k} - K_{2,k}P_{2,\tilde{Z}_{2,k}\tilde{Z}_{2,k}}K^T_{2,k} \tag{5.38}$$

$$K_{2,k} = P_{2,\tilde{x}_k\tilde{Z}_{2,k}} P^{-1}_{2,\tilde{Z}_{2,k}\tilde{Z}_{2,k}} \tag{5.39}$$

where $x^+_{2,k}$ and $P^+_{2,k}$ are results of static filter at time t_k.

5.1.2.3 Comparison Between Federated Filter and Kinematic and Static Filter

The spacecraft dynamic model information is used only once in the one state resolution of the kinematic and static filter. This effectively avoids the repetitive use of information and ensures the global optimality of final information fusion. However, the parallel structure of federated filter enables parallel calculation and saves the computation time. The kinematic and static filter is a cascaded filter, and the static filter does not work unless it receives the result of the kinematic filter, so the parallel calculation is impossible for it. On the other hand, for XNAV/CNS integrated navigation system, the observation period of X-ray pulsar is much longer than the sampling period of traditional celestial body measurement information. The positioning of spacecraft works by the celestial body measurement information at most of time. The effectiveness of parallel calculation is not obvious in this case.

In addition, in order to cope with the problems of deep space navigation mission, e.g. large scale of the state and navigation step, high uncertainty of initial navigation state, etc., the unscented kinematic and static filter can be further modified into a squared root form, which will not be discussed here for simplicity.

5.1.3 Error Compensation Method Based on Error Separation Principle

According to the analysis in Chap. 6, limited to current astronomical observation level, the X-ray pulsar-based navigation system has many system errors. If such system errors are not handled, it is hard to ensure the performance of the integrated

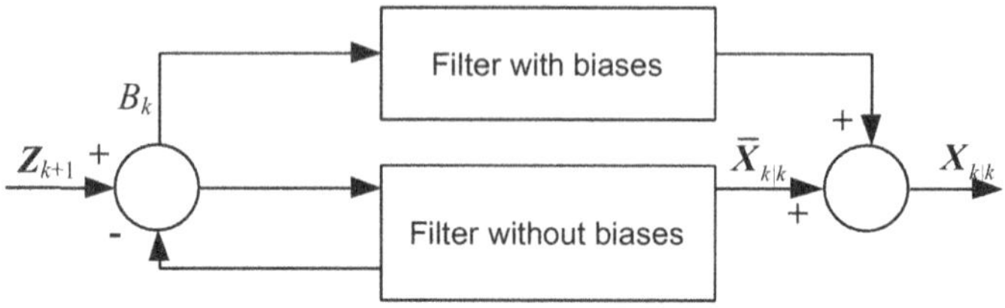

Fig. 5.5 Block diagram of the error separation method

navigation system. Considering that the autonomous navigation mission of Earth-orbiting orbit spacecraft is required to have an accuracy of better than 100 m, the positioning accuracy of deep space explorer is required to be the order of kilometer, and the influence of system errors is basically less than 1 km, only the system error compensation for integrated navigation system of Earth orbiting spacecraft is discussed in this subsection.

Based on the analysis in Chap. 6, the pulsar angular position error, ephemeris error of the Earth, pulsar distance error and satellite-borne atomic clock offset for Earth orbiting spacecraft can all be modeled as a slow time-varying process. Therefore, the augmented state method and the epoch-differenced method can be adopted for system error compensation.

To avoid the potential instability of numeric calculation in the augmented state method, the error separation method is utilized to handle the impact of system error in this section.

The error separation method is a parallel filtering algorithm proposed by Friedland in 1969 [10]. Its core principle is decoupling of the extended filter into two parallel filters. The first filter is called "filter without biases" which works based on the assumption that no systemstic biases exist. The second one is called "filter with biases" which is used for estimating the biases. Results of these two filters will be finally reorganized to generate the state of the original system. The block diagram of the error separation method is illustrated in Fig. 5.5.

If the kinematic and static filter is used for integrated navigation, the error separation can be enabled as soon as the structure of static filter is modified considering that the X-ray pulsar measurement data is processed only by the static filter.

In the static filter, the impact of system error $\boldsymbol{B}_p$ is neglected. Therefore, this filter can be used as a "filter without biases". The filter with biases and its final result are shown as below.

Assume that the predicted systematic biases and the corresponding error variance matrix at the time t_{k-1} is given by

$$\hat{\boldsymbol{B}}_{p,k-1} = E(\boldsymbol{B}_{p,k-1})\ \boldsymbol{P}_{k-1} = E\Big[(\boldsymbol{B}_{p,k-1} - \hat{\boldsymbol{B}}_{p,k-1})(\boldsymbol{B}_{p,k-1} - \hat{\boldsymbol{B}}_{p,k-1})^T\Big]. \quad (5.40)$$

The systematic bias estimation is given by

$$\boldsymbol{B}_{p,k}^{-} = \hat{\boldsymbol{B}}_{p,k-1}\ \boldsymbol{P}_{B,k}^{-} = \hat{\boldsymbol{P}}_{B,k-1} + \boldsymbol{Q}_{B,k-1} \quad (5.41)$$

$$P_{B,\tilde{x}_k\tilde{Z}_{2,k}} = P_{B,k}^- S^T \; P_{B,\tilde{Z}_{2,k}\tilde{Z}_{2,k}} = SP_{B,k}^- S^T + P_{p,\tilde{Z}_{2,k}\tilde{Z}_{2,k}} \tag{5.42}$$

$$\begin{aligned}\hat{B}_k &= B_k^- + K_{B,k}\left(Z_{2,k} - \hat{Z}_{2,k}^- - SB_k^-\right)\hat{P}_{B,k} \\ &= P_{B,k}^- - K_{B,k}P_{B,\tilde{Z}_{2,k}\tilde{Z}_{2,k}}K_{B,k}^T\end{aligned} \tag{5.43}$$

$$K_{B,k} = P_{B,\tilde{x}_k\tilde{Z}_{2,k}} P_{B,\tilde{Z}_{2,k}\tilde{Z}_{2,k}}^{-1}. \tag{5.44}$$

Then, the final estimation at the time t_k is given by

$$\hat{x}_k = x_{2,k}^+ + V_k\hat{B}_{p,k} \; \hat{P}_k = P_{p,k}^+ + V_k\hat{P}_{B,k}V_k^T. \tag{5.45}$$

In Eqs. (5.41)–(5.45),

$$S_k = H_kU_k + I \; V_k = U_k - K_{2,k}S_k \tag{5.46}$$

$$U_k = V_{k-1} \; H_k = \left.\frac{\partial h_2(x)}{\partial x}\right|_{x=x_{2,k}^-} \tag{5.47}$$

where I is the unit matrix.

5.1.4 Simulation Analysis

5.1.4.1 Simulation of XNAV/Stellar Angle Measurement Integrated Navigation System

The autonomous navigation mission of high Earth orbit satellites is used for analyzing the performance of XNAV/stellar angle measurement integrated navigation system. The orbits of three satellites, NTS 2, DOGE 1 and ATS 1 are analyzed. In the J2000.0 Earth-centered inertial system, the initial orbital elements of these satellites are as shown in Table 5.1. Pulsars selected by the Microcosm Company are taken as the navigation pulsars. The 50 stars distributed over the celestial sphere are considered as the reference stars (with brightness ≤ 2 m).

The pulsars, PSR B0531+21, PSR B1821−24 and PSR B1509−58, are selected as the navigation pulsars. Assume that the simulation duration is 7 d, the angular position error of pulsars is 0.001″, the distance error is 30% and the satellite-borne atomic clock offset is 1 μs. DE405 is used for providing the predicted Earth position during the navigation, and the difference between DE421and DE405 is used for simulating the ephemeris error of the Earth.

The standard deviation of the measurement of X-ray pulsar is determined by Eq. (5.3), where the area of the detector is 0.3 m^2, the time resolution of the detector is 1 μs and the X-ray background radiation flux is 50 ph/m^2/s. The observation period

Table 5.1 Initial orbital elements of satellites

Orbital elements	Satellite name		
	NTS 2	DODGE 1	ATS 1
Semi-major axis [km]	26,570.08	39,842.89	42,150.06
Eccentricity	0.0047	0.0051	0.0017
Inclination [°]	63.15	10.11	9.2
Right ascension of ascending node [°]	136.31	330.92	314.83
Argument of perigee [°]	211.64	177.7	155.1
Mean anomaly [°]	304.82	121.27	319.79

of pulsar is 1800 s, and the sampling period of stellar angle measurement is 300 s. The accuracy of star sensor is $3''$ (1σ), and the accuracy of horizon sensor is $0.05°$ (1σ).

Figure 5.6 compares the performances of the integrated navigation system, the X-ray pulsar-based navigation system and the navigation system based on stellar angle measurement. As shown in Fig. 5.6, all three methods are convergent, but the integrated navigation system has a better steady state. Table 5.2 gives the navigation

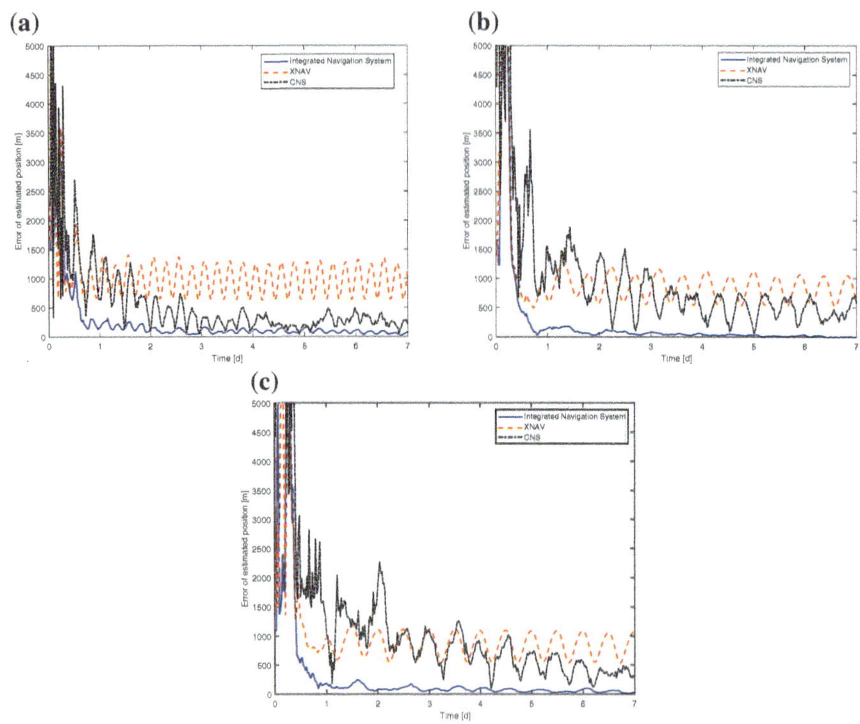

Fig. 5.6 Performance comparison of the integrated navigation system, X-ray pulsar-based navigation system and navigation system based on stellar angle distance

Table 5.2 Positioning accuracy comparison of three navigation systems [m]

Navigation system	Satellite name		
	NTS 2	DODGE 1	ATS 1
Integrated navigation system	86.02	64.99	76.72
XNAV	994.52	827.93	845.1
Navigation system based on stellar angle	273.27	367.48	461.62

accuracy comparison obtained from 300 Monte Carlo simulations. For these satellites, the positioning accuracy of the integrated navigation system is less than 100 m, but that of the other two is larger than 200 m. Therefore, the performance of integrated navigation system is superior to XNAV and the navigation system based on stellar angle measurement. This is mainly because the integrated navigation system integrates the measurement information of X-ray pulsars and the stellar angle measurement. Moreover, from Fig. 5.6 and Table 5.2, we can see that the performance of the integrated navigation system is little affected by the orbit altitude, while that of the navigation system based on the stellar angle measurement is obviously affected by the orbit altitude. Therefore, the former has an application scope broader than the latter one.

As the positioning result for NTS 2 is the worst, we will focus on analyzing the performance of integrated navigation system by using NTS 2 in the following part.

Figure 5.7 shows the positioning estimation error and the corresponding 3σ profile. The positioning estimation error of the integrated navigation system is within the 3σ profile in the whole process, but that of the X-ray pulsar-based navigation system is out. This indicates that the integrated navigation system has compensated most system error of XNAV.

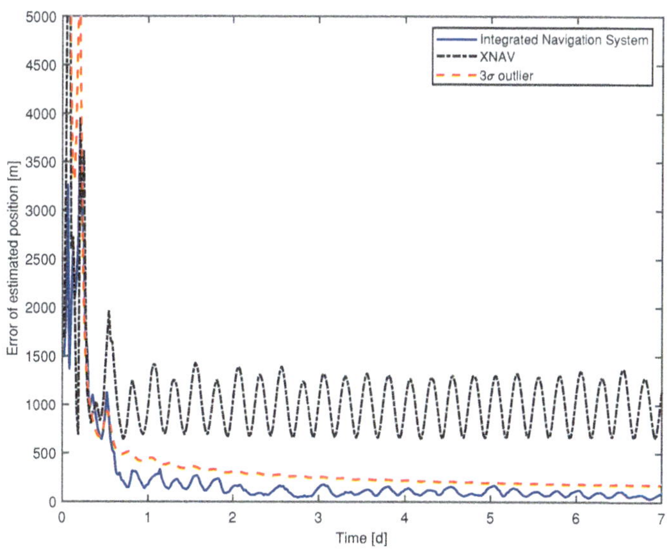

Fig. 5.7 System error compensation performance of the integrated navigation system

Table 5.3 Initial orbital elements of deep space explorer

Orbital Elements	Value
Semi-major axis [km]	193,217,604.27
Eccentricity	0.2364
Inclination [°]	23.45
Right ascension of ascending node [°]	0.2579
Argument of perigee [°]	71.35
Mean anomaly [°]	117.05

5.1.4.2 Simulation of XNAV/Sun Information Integrated Navigation System

The performance of X-ray pulsar/Sun information integrated navigation system is analyzed by designing an autonomous navigation mission of deep space explorer.

Assume that a deep space explorer is flying from the Earth to Mars. In the Sun-centered inertial system, the initial orbital elements of the deep space explorer are as shown in Table 5.3. PSR B0531+21, PSR B1821−24 and PSR B1509−58 are selected as the navigation pulsars. The spacecraft only carries a sensor with a pointing adjustment device, used for observing different pulsars. The area of the sensor is 0.3 m^2. The simulation duration is 50 d, the observation period of pulsars is 2 h, and the pulsar angular position error is 0.002″. The accuracy of Sun sensor is 0.001° (1σ), and the accuracy of spectrometer is 0.01 m/s (1σ), both having a sampling period of 5 min. The initial navigation errors of the deep space explorer are [100 km, 100 km, 100 km, 10 m/s, 10 m/s, 10 m/s].

Figure 5.8 shows the integrated navigation system, XNAV, the Sun information navigation system and the modified Sun information navigation system described in [4]. Besides the Sun direction vector and the radial Doppler speed measurements, the Earth direction vector measurement made by the deep space explorer is also adopted for the modified Sun information navigation system. Similarly, the measurement accuracy of Earth direction vector is taken as 0.001° (1σ), and the sampling period is 5 min. As can be seen in Fig. 5.8, all the methods converge. Therefore, they are all applicable to the autonomous navigation mission of deep space explorer. However, it would take more than 10 days for the Sun information navigation system and the modified Sun information navigation system to converge. In contrast, the convergence of integrated navigation system and XNAV can be completed within 5 days. Table 5.4 lists the navigation accuracy obtained from 300 Monte Carlo simulations. Results show that the performance of the integrated navigation system is superior to the other three.

The necessity of introducing the Sun information into XNAV is analyzed below. Figure 5.9 illustrates the accumulation of navigation error in the first observation period of pulsar. Despite of the use of Sun information, the navigation error increases as usual, but its increasing speed is far slower than XNAV. Therefore, the use of Sun information will decrease the navigation error accumulating within the observation periods of pulsar and enhance the convergence speed.

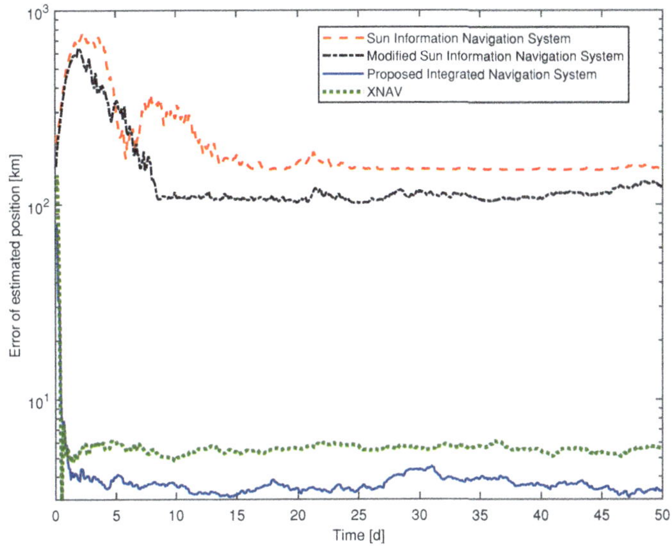

Fig. 5.8 Performance comparison of the integrated navigation system, Sun information navigation system, modified Sun information navigation system and XNAV

Table 5.4 Positioning accuracy comparison of four navigation systems

Navigation system	Integrated navigation system	XNAV	Modified Sun information navigation system	Sun information navigation system
Position estimation accuracy [km]	3.37	5.47	123.82	155.59

Figure 5.10 compares the performances of two X-ray pulsar observation schemes. In scheme 1, the deep space explorer keeps observing the pulsar PSR 0531+21; in scheme 2, the deep space explorer observes three selected pulsars in sequence. As shown in Fig. 5.10, although both the schemes could ensure filters converge, the performance of scheme 2 is better. This is because the geometrical structure in scheme 2 is better than that in method 1. In order to show the difference in geometrical structure, Table 5.5 gives the average observability degree (condition number of observable matrix) of this two schemes. As a result, the observability degree of scheme 2 is superior to scheme 1. Therefore, although the integrated navigation system fuses the X-ray pulsar and the Sun information, the geometrical structure of X-ray pulsar observation has an influence on the final navigation result, so the deep space explorer shall observe sequentially different pulsars.

Figure 5.11 gives the variation of navigation performance with respect to the orbital semi-major axis of the deep space explorer. The performances of integrated system and Sun information navigation system degrade with the increase of orbital semi-major axis. Although the semi-major axis of deep space explorer reaches 3 AU,

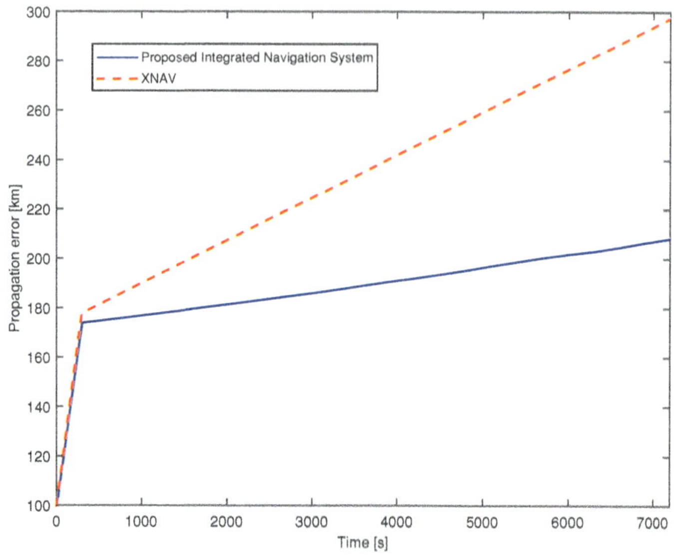

Fig. 5.9 Navigation error accumulation in the first observation period of X-ray pulsar

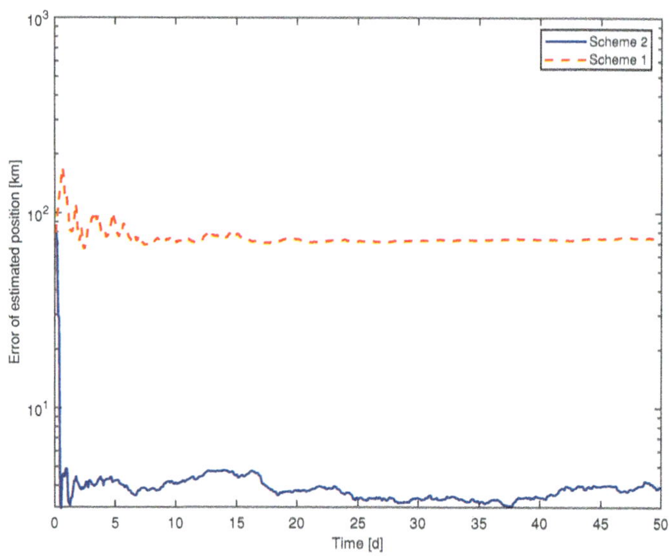

Fig. 5.10 Performance comparison between two X-ray pulsar observation schemes

Table 5.5 Average observability degree of two schemes

Scheme	Observability degree (condition number of observable matrix)
Scheme 1	1328.88
Scheme 2	1173.31

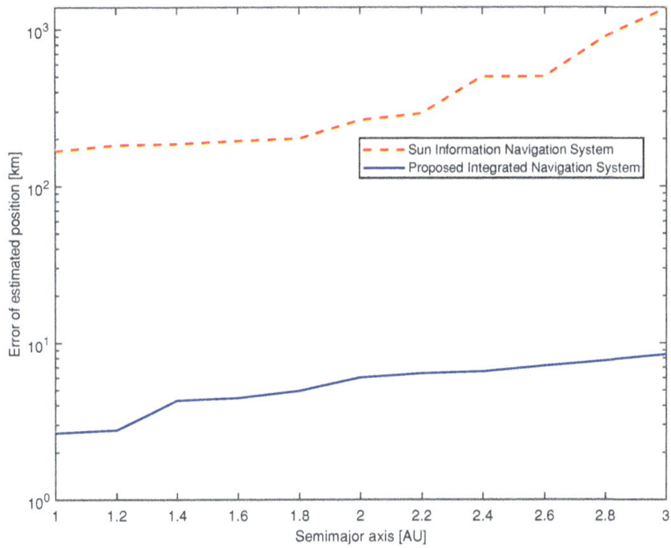

Fig. 5.11 Variation of navigation performance with the orbital semi-major axis of deep space explorer

the positioning accuracy of integrated navigation system is still superior to 10 km. In the same case, the positioning accuracy of Sun information integrated navigation system is larger than 1000 km. Therefore, the use of integrated navigation system might mitigate the impact of the orbital semi-major axis variation of deep space explorer on the navigation performance, so it has an application scope broader than the Sun information navigation system.

5.2 XNAV/INS Integrated Navigation Framework

Being of a high specific impulse, the low-thrust system can perform many challenging space missions and has aroused global attention [11]. At present, the research on low-thrust system mostly focuses on the design of low-thrust system and orbit, and there are rare researches on autonomous navigation systems that are applicable to low-thrust deep space exploration [12, 13]. It is necessary to study autonomous navigation methods to improve the autonomy and viability of low-thrust deep space explorer.

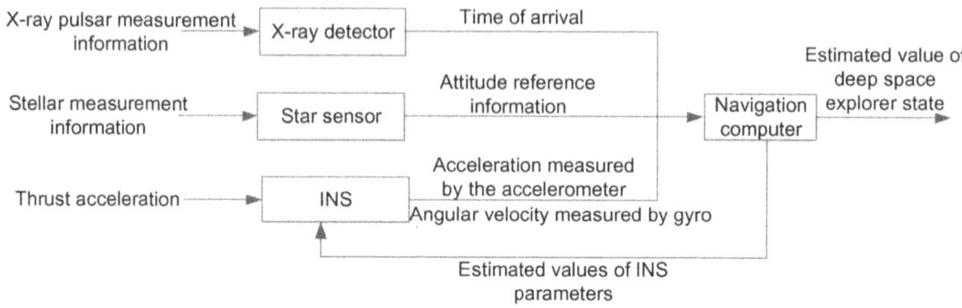

Fig. 5.12 Block diagram of XNAV/INS integrated navigation system

Garulli et al. proposed the autonomous navigation system that is applicable to Earth orbiting spacecraft [14]. This system is composed of an INS, a star tracker and a GPS receiver, and its positioning accuracy is higher than 20 m. However, GPS is not suitable for deep space explorer due to its rather weak signal and visibility. Friedlander designed an autonomous navigation system that is applicable to interplanetary low-thrust spacecraft [15]. This navigation system is composed of an INS, a star sensor and a planet tracker, and the position and attitude information of spacecraft are determined respectively by using stellar angle measurement and observing the target star. As analyzed in Sect. 5.1, the positioning accuracy of this method is only about thousand kilometers.

Inspired by the method of Friedlander, this section proposes an autonomous navigation method of low-thrust deep space explorer based on the X-ray pulsar/INS. In this method, INS can provide the real-time position and attitude information of the deep space explorer, and the X-ray pulsar measurement information is introduced into correcting the long-term system error accumulation of INS.

5.2.1 Composition of XNAV/INS Integrated Navigation System

This navigation system is composed of an INS, three X-ray sensors and two star sensors. Among them, the X-ray explorer detects the position measurement information, while the star sensors provide the attitude determination information. The measurement information of X-ray detectors and star sensors fuse with the output of INS in order to mitigate the long-term system error drift of the gyro and accelerometer. The basic structure of such system is shown in Fig. 5.12.

5.2.2 Dynamic Model

The state of this navigation system has 16 dimensions, including the position and velocity vector of deep space explorer in the heliocentric inertial system, the quaternion representing the spacecraft attitude and errors of the gyro and accelerometer defined in the deep space exploration system. The dynamic model is [16]

$$\dot{\boldsymbol{r}} = \boldsymbol{v} + \boldsymbol{w}_{\boldsymbol{r}} \tag{5.48}$$

$$\dot{\boldsymbol{v}} = -\frac{\mu_S}{\|\boldsymbol{r}\|^3}\boldsymbol{r} + \sum_{j=1}^{n_p} \mu_j \left(\frac{\boldsymbol{r}_{rj}}{\|\boldsymbol{r}_{rj}\|^3} - \frac{\boldsymbol{r}_{pj}}{\|\boldsymbol{r}_{pj}\|^3} \right) + \boldsymbol{a}_{H.O.T} + \boldsymbol{R}_b^i \cdot \boldsymbol{f}^b + \boldsymbol{w}_v \tag{5.49}$$

$$\dot{\boldsymbol{q}}_b^i = \frac{1}{2}\begin{bmatrix} \boldsymbol{\omega}_b \\ 0 \end{bmatrix} \otimes \boldsymbol{q}_b^i + \boldsymbol{w}_q \tag{5.50}$$

$$\dot{\boldsymbol{D}} = \frac{1}{\alpha}\boldsymbol{D} + \boldsymbol{w}_d \tag{5.51}$$

$$\dot{\boldsymbol{B}} = \frac{1}{\beta}\boldsymbol{B} + \boldsymbol{w}_b \tag{5.52}$$

where $\boldsymbol{r}$ and $\boldsymbol{v}$ are position and velocity of the spacecraft, $\boldsymbol{R}_b^i$ is the attitude matrix of the deep space explorer relative to the inertial system, $\boldsymbol{f}^b$ is the specific force measured by the accelerometer, $\boldsymbol{q}_b^i$ is the attitude quaternion, and $\boldsymbol{\omega}_b$ is the angular velocity measured by the gyro. The system errors, $\boldsymbol{D}$ and $\boldsymbol{B}$, of gyro and accelerometer are modeled as one-order Markov processes, and α and β are corresponding inverted correlation coefficients. The meaning of other variables in Eq. (5.49) is the same as Eq. (2.35).

In Eqs. (5.48)–(5.52), $\boldsymbol{w}_{\boldsymbol{r}}$, $\boldsymbol{w}_v$, $\boldsymbol{w}_q$, $\boldsymbol{w}_d$ and $\boldsymbol{w}_b$ are process noises.

5.2.3 Observation Model

The observation models of this integrated navigation system include the X-ray pulsar-based navigation model and the attitude measurement model for star sensor. The former is the same as Eq. (4.13) and the latter can be expressed as

$$\begin{bmatrix} \boldsymbol{y}_{A1} \\ \boldsymbol{y}_{A2} \end{bmatrix} = \begin{bmatrix} \left(\boldsymbol{R}_b^i(\boldsymbol{q}_b^i)\right)^{-1}\boldsymbol{n}^1 \\ \left(\boldsymbol{R}_b^i(\boldsymbol{q}_b^i)\right)^{-1}\boldsymbol{n}^2 \end{bmatrix} + \boldsymbol{\eta} \tag{5.53}$$

where $\boldsymbol{n}^1$ and $\boldsymbol{n}^2$ are the direction vectors of the star observed by the star sensor in the inertial system, $\boldsymbol{y}_{A1}$ and $\boldsymbol{y}_{A2}$ are star direction vectors measured by the deep space explorer in its system, and $\boldsymbol{\eta}$ is the zero-mean Gaussian white noise.

Table 5.6 Constraints of the low-thrust orbit

Upper limit of thrust [N]	Specific impulse [s]	Initial mass of deep space explorer [kg]
0.39	3000	1000

Table 5.7 INS parameters

Parameters	Gyro	Accelerometer
Initial error [rad/s]	5×10^{-7}	1×10^{-4}
Inverted correlation coefficient [s]	3600	3000
Standard variance of $\boldsymbol{w}_d$ [rad/s^2]	1×10^{-8}	1×10^{-5}

5.2.4 Simulation Analysis

Assume that a low-thrust deep space explorer is flying from the Earth to Mars. The low-thrust orbit is obtained by using Gauss pseudo-spectral method [17]. Constraints of the low-thrust orbit are shown in Table 5.6.

Assume that the deep space explorer has left the sphere of influence of Earth, so GPS cannot be used. The initial states of the deep space explorer are [−145,951,721.41 km, 57,648,355.77 km, 249,070 km] and [−5.95 km/s, −31.21 km/s, 0.16 km/s]. The initial navigation errors are [100 km, 100 km, 100 km] and [10 m/s, 10 m/s, 10 m/s]. The effective areas of three X-ray detectors are 0.3 m^2. Suppose that the deep space explorer can observe three pulsars, PSR B0531+21, PSR B1821−24 and PSR B1509−58, at the same time. The observation period of X-ray pulsar is 1 h, and that for the attitude determination of the star sensor is 2 s. The standard deviation of X-ray pulsar positioning observation noise is determined by Eq. (4.15). Assume that a poor-quality star sensor is used, and take its angle measurement accuracy as 0.05° (1σ). The output of INS is continuous and parameters of INS are shown in Table 5.7.

Figures 5.13 and 5.14 show the performance of this integrated navigation system, with the full line indicating the estimated error and dotted line indicating the 3σ profile. The estimated errors of position and attitude are convergent within the 3σ profile, which indicates the estimated result of this integrated navigation system is unbiased. The positioning result of the Y-axis are much superior to that of the X-axis and Z-axis, which is involved with the geometrical structures of the deep space explorer and pulsars. It can be seen from Fig. 5.14, the attitude determination accuracy of the integrated navigation system can be superior to 0.1°. Therefore, the XNAV/INS integrated navigation system depicted in this section can provide the position and attitude information of the low-thrust deep space explorer at the same time.

Figures 5.15 and 5.16 show the performance comparison of the XNAV/INS integrated navigation system, INS and Friedlander method. For the Friedlander method, the stellar angle measurement and the stellar attitude determination information are

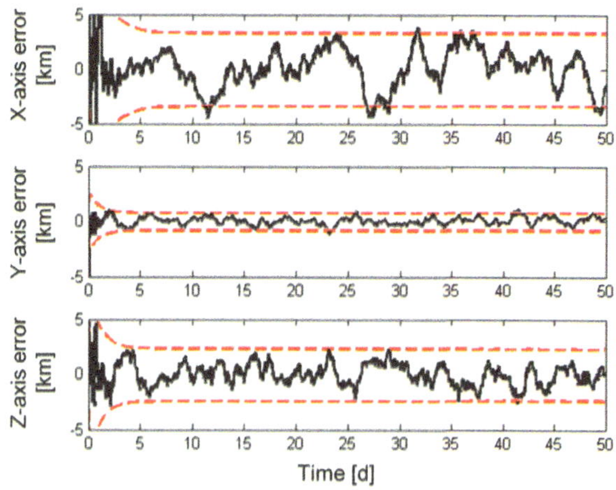

Fig. 5.13 Position estimation error of XNAV/INS integrated navigation system

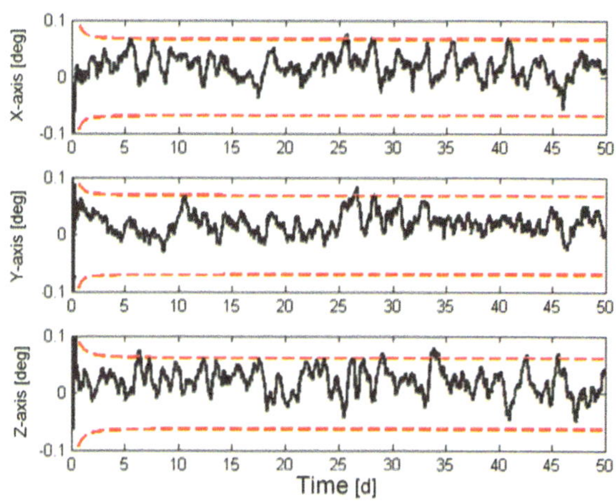

Fig. 5.14 Attitude estimation error of XNAV/INS integrated navigation system

assumed to be generated by two star sensors with the accuracy of 3″ (1σ) and one optical camera with the accuracy of 0.5° (1σ).

It can be seen from Fig. 5.15 that the position estimation error curve of INS diverges over time, while the curves for XNAV/INS integrated navigation system and Friedlander method are convergent. However, the XNAV/INS integrated navigation system would converge faster than Friedlander method, and its final positioning accuracy is superior to 10 km. Figure 5.16 indicates that the attitude estimation error curve of INS also diverges, but the convergence of attitude estimation errors will be ensured by other two methods. The attitude determination accuracy of XNAV/INS integrated navigation system is worse than that of the Friedlander method. This is because the star sensor adopted in the integrated navigation system is inferior to the star sensor adopted in the Friedlander method.

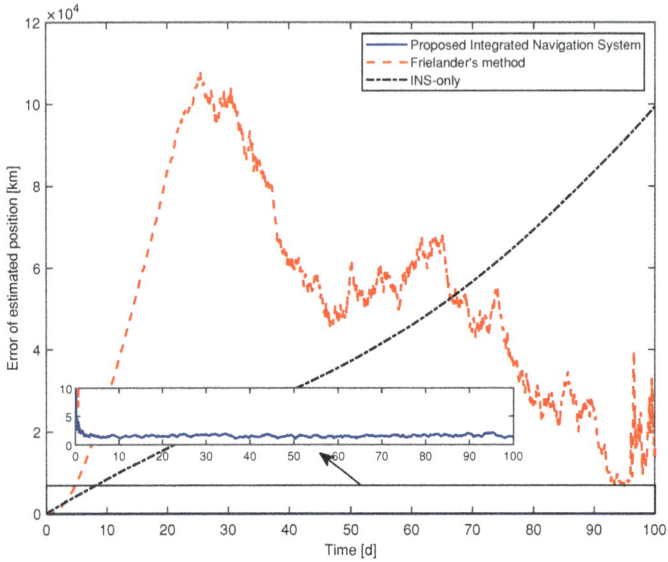

Fig. 5.15 Positioning performance comparison

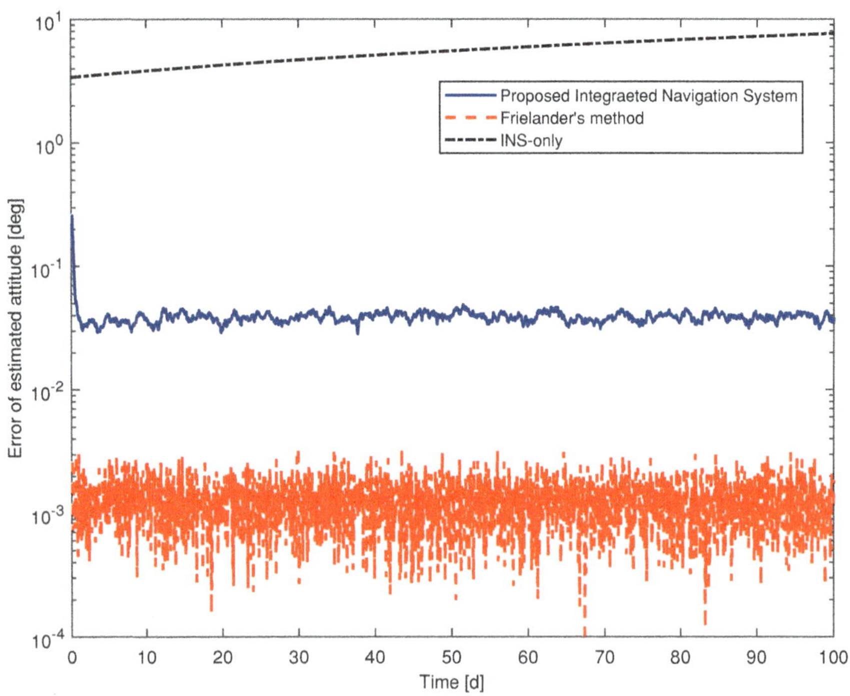

Fig. 5.16 Comparison of the attitude determination accuracy

5.3 Summary

In order to compensate the deficiencies of XNAV, this chapter proposes two X-ray pulsar/multiple measurement information fused navigation frameworks. For the XNAV/CNS integrated navigation method, we emphatically introduce the XNAV/stellar angle measurement integrated navigation system and the XNAV/Sun information integrated navigation system, analyze the optimum information fusion method and investigate the performance of these two integrated navigation systems applying for the high Earth orbit satellite and deep space explorer respectively. For the XNAV/INS integrated navigation method, we analyze the effect that the information of X-ray pulsars and of star sensor could reduce the long-term drift of INS.

There are a variety of integrated navigation forms, and it is hard to describe them all in this section. However, based on the integrated navigation frameworks provided in this chapter, different integrated navigation methods can be derived. For example, XNAV/ radar ranging integrated navigation method, XNAV/ ultraviolet imaging integrated navigation method can be derived from the XNAV/CNS integrated navigation framework. A series of integrated navigation systems suitable for maneuvering spacecraft can be derived by fusing different celestial body measurement information to the fundamental XNAV/INS integrated navigation framework, which will not be discussed in detail here.

References

1. Sheikh SI (2005) The use of variable celestial X-ray sources for spacecraft navigation. University of Maryland, Maryland
2. Fang J, Ning X (2006) Principle and application of celestial navigation. Beihang University Press, Beijing
3. Guo YP (1999) Self-contained autonomous navigation system for deep space missions. Adv Astronaut Sci 102:1099–1113
4. Yim JR, Crassidis JL, Junkins JL (2000) Autonomous orbit navigation of interplanetary spacecraft. AIAA guidance, navigation, and control conference, Denver, USA
5. Carlson AA (1988) Federated filter for fault-tolerant integrated navigation system. In: Position location and navigation symposium, Orlando, FL
6. Yang Y (2006) Adaptive dynamic navigation and position. Surveying and Mapping Press, Beijing
7. Yang Y (2003) Multi-source dynamic and static filter integrated navigation. J Wuhan Univ (Inf Sci Ed) 28(4):386–388
8. Yang YX, Cui X, Gao WG (2004) Adaptive integrated navigation for multi-sensor adjustment outputs. J Navig 57:287–285
9. Wang YD, Zheng W, An XY et al (2013) XNAV/CNS integrated navigation based on improved kinematic and static filter. J Navig 66:899–918
10. Friedland B (1969) Treatment of bias in recursive filtering. IEEE Trans Autom Control AC-14:359–367
11. Rayman M, Fraschetti T, Raymond C (2004) A mission in development for exploration of main belt asteroids vesta and ceres. In: 55th international astronautical congress, Vancouver, Canada
12. Patel P, Scheeres D, Gallimore A (2006) Maximizing payload mass fractions of spacecraft for interplanetary electric propulsion missions. J Spacecr Rocket 43:822–827

13. Brophy J (2003) Advanced ion propulsion systems for affordable deep-space missions. Acta Astronaut 52:309–316
14. Garulli A, Giannitrapani A (2011) Autonomous low-earth-orbit station-keeping with electric propulsion. J Guid Control Dyn 34:1683–1693
15. Friedlander A (1966) Analysis of an optimal celestial-inertial navigation concept for low-thrust interplanetary vehicle. NASA report
16. Wen Y (2012) Research on navigation method of high orbit auto-transfer spacecraft. National University of Defense Technology
17. Tang G, Zhang H, Zhen W, Wang G (2013) Low-thrust orbital maneuver dynamic and control. Science Press, Beijing

Chapter 6
Spacecraft Autonomous Navigation Using the X-Ray Pulsar Time Difference of Arrival

6.1 Shortcomings of Autonomous Navigation Using Inter-satellite Link

6.1.1 Inter-satellite Link Ranging Measurement

Supposing the spacecraft i and spacecraft j receive the signal from each other at time t_0, the signal received by spacecraft i was emitted by spacecraft j at time t_b, and the signal received by spacecraft j was emitted by spacecraft i at time t_a. Taking the errors within the measured time into consideration, t_a and t_b can be described by Eqs. (6.1) and (6.2) where t_a^* and t_b^* are the true values of t_a and t_b, ε_a and ε_b are the uncertainties and $\delta_i(t)$ and $\delta_j(t)$ are the systematic biases within (Fig. 6.1).

For spacecraft i

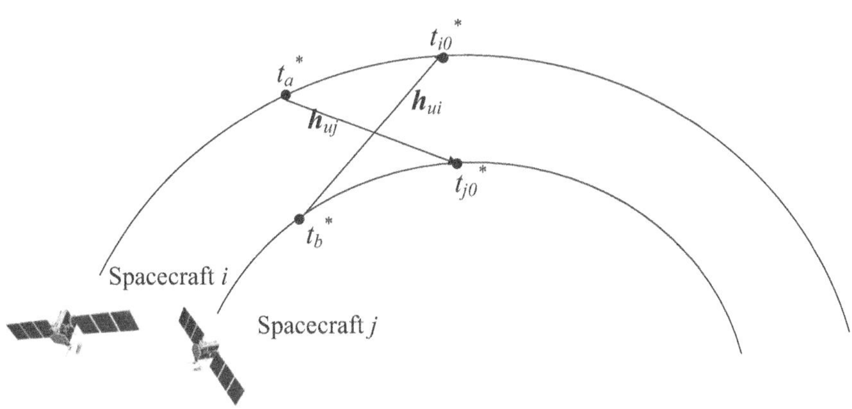

Fig. 6.1 Navigation based on inter-satellite link

W. Zheng and Y. Wang, *X-ray Pulsar-based Navigation*, Navigation: Science and Technology 5, https://doi.org/10.1007/978-981-15-3293-1_6

$$\begin{aligned} t_0 &= t_{i0}^* + \delta_i\left(t_{i0}^*\right) + \varepsilon_{i0} \\ t_a &= t_a^* + \delta_i\left(t_a^*\right) + \varepsilon_a \\ \delta_i &= const \approx \delta_i\left(t_{i0}^*\right) \approx \delta_i\left(t_a^*\right) \end{aligned} . \tag{6.1}$$

For spacecraft j

$$\begin{aligned} t_0 &= t_{j0}^* + \delta_j\left(t_{j0}^*\right) + \varepsilon_{j0} \\ t_b &= t_b^* + \delta_j\left(t_b^*\right) + \varepsilon_b \\ \delta_j &= const \approx \delta_j\left(t_{j0}^*\right) \approx \delta_j\left(t_b^*\right) \end{aligned} . \tag{6.2}$$

The measurement z_{ui} at spacecraft i is

$$\begin{aligned} z_{ui} &= \left|\boldsymbol{r}_i\left(t_{i0}\right) - \boldsymbol{r}_j\left(t_b\right)\right| = \boldsymbol{h}_{ui} + c \cdot \left(\delta_i - \delta_j\right) + c \cdot \left(\varepsilon_{i0} - \varepsilon_b\right) + \varepsilon_i \\ \boldsymbol{h}_{ui} &= \left|\boldsymbol{r}_i\left(t_{i0}^*\right) - \boldsymbol{r}_j\left(t_b^*\right)\right| \end{aligned} \tag{6.3}$$

where ε_i is the measurement error. Assuming $\Delta_{ui} = c \cdot (\varepsilon_{i0} - \varepsilon_b) + \varepsilon_i$, we have

$$z_{ui} = \boldsymbol{h}_{ui} + c \cdot \left(\delta_i - \delta_j\right) + \Delta_{ui}. \tag{6.4}$$

In the same way, the measurement z_{uj} at spacecraft j can be expressed as

$$z_{uj} = \boldsymbol{h}_{uj} + c \cdot \left(\delta_j - \delta_i\right) + \Delta_{uj} \tag{6.5}$$

where

$$\boldsymbol{h}_{uj} = \left|\boldsymbol{r}_j\left(t_{j0}^*\right) - \boldsymbol{r}_i\left(t_a^*\right)\right| \tag{6.6}$$

$$\Delta_{uj} = c \cdot \left(\varepsilon_{j0} - \varepsilon_a\right) + \varepsilon_j \tag{6.7}$$

In order to reduce the impact of Δ_{ui} and of Δ_{uj} on the measurement, z_{ui} and z_{uj} can be combined as

$$z_u = \left(z_{ui} + z_{uj}\right)/2 = \boldsymbol{h}_u + \Delta_u \tag{6.8}$$

where

$$\begin{aligned} \boldsymbol{h}_u &= \left(\boldsymbol{h}_{ui} + \boldsymbol{h}_{uj}\right)/2 \\ &= \left(\left|\boldsymbol{r}_i\left(t_{i0}^*\right) - \boldsymbol{r}_j\left(t_b^*\right)\right| + \left|\boldsymbol{r}_j\left(t_{j0}^*\right) - \boldsymbol{r}_i\left(t_a^*\right)\right|\right)/2 \\ &= \left(\left|\boldsymbol{r}_i(t_0) - \boldsymbol{r}_j(t_0)\boldsymbol{\Phi}_j(t_b, t_0)\right| + \left|\boldsymbol{r}_j(t_0) - \boldsymbol{r}_i(t_0)\boldsymbol{\Phi}_i(t_a, t_0)\right|\right)/2 \end{aligned} \tag{6.9}$$

with $\boldsymbol{\Phi}_j(t_b, t_0)$ and $\boldsymbol{\Phi}_i(t_a, t_0)$ are the state transition matrices.

6.1.2 Mathematical Analysis for Orbit Determination Using Inter-satellite Link Ranging

For spacecraft i and spacecraft j, the inter-satellite link ranging measurement is

$$\boldsymbol{r}_{ij} = r_{ij}\,\boldsymbol{e}_{ij} = \boldsymbol{r}_j - \boldsymbol{r}_i \tag{6.10}$$

where $\boldsymbol{e}_{ij}$ is the unit vector which directs from spacecraft i to spacecraft j. Differentiating the above equation yields

$$\begin{aligned} d\boldsymbol{r}_{ij} &= r_{ij}d\boldsymbol{e}_{ij} + dr_{ij}\boldsymbol{e}_{ij} = d\boldsymbol{r}_j - d\boldsymbol{r}_i \\ &= \sum \frac{\partial \boldsymbol{r}_j(q_{j0}, t)}{\partial q_{j0}}\Delta q_{j0} - \sum \frac{\partial \boldsymbol{r}_i(q_{i0}, t)}{\partial q_{i0}}\Delta q_{i0} \end{aligned} \tag{6.11}$$

where q_{i0}, q_{j0} are the orbital elements of the two spacecraft at epoch time respectively.

Equation (6.11) is the condition equation of autonomous navigation using inter-satellite link ranging, A is the coefficient matrix of the condition equation, namely

$$\begin{aligned} A &= \left(-\left(\frac{\partial \boldsymbol{r}_i}{\partial a_i}, \frac{\partial \boldsymbol{r}_i}{\partial e_i}, \frac{\partial \boldsymbol{r}_i}{\partial i_i}, \frac{\partial \boldsymbol{r}_i}{\partial \Omega_i}, \frac{\partial \boldsymbol{r}_i}{\partial \omega_i}, \frac{\partial \boldsymbol{r}_i}{\partial M_i}\right), \frac{\partial \boldsymbol{r}_j}{\partial a_j}, \frac{\partial \boldsymbol{r}_j}{\partial e_j}, \frac{\partial \boldsymbol{r}_j}{\partial i_j}, \frac{\partial \boldsymbol{r}_j}{\partial \Omega_j}, \frac{\partial \boldsymbol{r}_j}{\partial \omega_j}, \frac{\partial \boldsymbol{r}_j}{\partial M_j}\right). \\ &= \left(A_{a_i}, A_{e_i}, A_{i_i}, A_{\Omega_i}, A_{\omega_i}, A_{M_i}, A_{a_j}, A_{e_j}, A_{i_j}, A_{\Omega_j}, A_{\omega_j}, A_{M_j}\right) \end{aligned} \tag{6.12}$$

If the system is observable, $A^{\mathrm{T}}A$ is of full rank, that is to say, the rank of A is 12. The remainder of this section is to discuss when A will become rank defect.

(1) **Orbit inclination** i

Given

$$\frac{\partial \boldsymbol{r}}{\partial i} = \boldsymbol{J}_N \times \boldsymbol{r} \tag{6.13}$$

where $\boldsymbol{J}_N = \left(\cos\Omega \; \sin\Omega \; 0\right)^{\mathrm{T}}$, we have

$$\boldsymbol{e}_{ij} \cdot \frac{\partial \boldsymbol{r}_j}{\partial i_j} = -\frac{1}{r_{ij}}\left[\cos\Omega_j\left(\boldsymbol{r}_j \times \boldsymbol{r}_i\right)\cdot \boldsymbol{i} + \sin\Omega_j\left(\boldsymbol{r}_j \times \boldsymbol{r}_i\right)\cdot \boldsymbol{j}\right] \tag{6.14}$$

$$\boldsymbol{e}_{ij} \cdot \frac{\partial \boldsymbol{r}_i}{\partial i_i} = -\frac{1}{r_{ij}}\left[\cos\Omega_i\left(\boldsymbol{r}_i \times \boldsymbol{r}_j\right)\cdot \boldsymbol{i} + \sin\Omega_i\left(\boldsymbol{r}_i \times \boldsymbol{r}_j\right)\cdot \boldsymbol{j}\right] \tag{6.15}$$

where $\boldsymbol{i}, \boldsymbol{j}$ are the unit vector of X, Y respectively, from Eqs. (6.14) and (6.15), it could be seen that when $\Omega_i = \Omega_j$,

$$e_{ij} \cdot \frac{\partial r_j}{\partial i_j} = e_{ij} \cdot \frac{\partial r_i}{\partial i_i} \tag{6.16}$$

and when $\Omega_i = \Omega_j + \pi$,

$$e_{ij} \cdot \frac{\partial r_j}{\partial i_j} = -e_{ij} \cdot \frac{\partial r_i}{\partial i_i}. \tag{6.17}$$

The coefficients of Δi_i and Δi_j in the condition equation are the same or opposite, thus A is rank defect.

(2) **Right ascension of ascending node** Ω

Given

$$\frac{\partial \boldsymbol{r}}{\partial \Omega} = \boldsymbol{J}_Z \times \boldsymbol{r} \tag{6.18}$$

where $\boldsymbol{J}_Z = \begin{bmatrix} 0 \; 0 \; 1 \end{bmatrix}^{\mathrm{T}}$, we have

$$e_{ij} \cdot \frac{\partial r_j}{\partial \Omega_j} = -\frac{1}{r_{ij}}(\boldsymbol{r}_j - \boldsymbol{r}_i) \cdot (\boldsymbol{J} \times \boldsymbol{r}_j) = -\frac{1}{r_{ij}}(\boldsymbol{r}_j \times \boldsymbol{r}_i) \cdot \boldsymbol{k} \tag{6.19}$$

$$e_{ij} \cdot \frac{\partial r_i}{\partial \Omega_i} = -\frac{1}{r_{ij}}(\boldsymbol{r}_j - \boldsymbol{r}_i) \cdot (\boldsymbol{J} \times \boldsymbol{r}_i) = -\frac{1}{r_{ij}}(\boldsymbol{r}_j \times \boldsymbol{r}_i) \cdot \boldsymbol{k}. \tag{6.20}$$

where $\boldsymbol{k}$ is the unit vector of Z direction, thus the following equation can be derived

$$e_{ij} \cdot \frac{\partial r_j}{\partial \Omega_j} = e_{ij} \cdot \frac{\partial r_i}{\partial \Omega_i}. \tag{6.21}$$

The coefficients of $\Delta\Omega_i$ and $\Delta\Omega_j$ in the condition equation are the same and thus A is rank defect.

(3) **Argument of perigee** ω

Given

$$\frac{\partial \boldsymbol{r}}{\partial \omega} = \boldsymbol{R} \times \boldsymbol{r}, \tag{6.22}$$

where $\boldsymbol{R}$ is the unit vector of the normal direction of the orbit with the expression of

$$\boldsymbol{R} = \frac{1}{\sqrt{\mu p}}(\boldsymbol{r} \times \dot{\boldsymbol{r}}). \tag{6.23}$$

In Eq. (6.22), p is the semi-latus rectum of the orbit, thus

$$e_{ij} \cdot \frac{\partial r_j}{\partial \omega_j} = -\frac{1}{r_{ij}}(\boldsymbol{r}_j - \boldsymbol{r}_i) \cdot (\boldsymbol{R}_j \times \boldsymbol{r}_j) = -\frac{1}{r_{ij}}(\boldsymbol{r}_j \times \boldsymbol{r}_i) \cdot \boldsymbol{R}_j \tag{6.24}$$

$$e_{ij} \cdot \frac{\partial r_i}{\partial \omega_i} = -\frac{1}{r_{ij}}(\boldsymbol{r}_j - \boldsymbol{r}_i) \cdot (\boldsymbol{R}_i \times \boldsymbol{r}_i) = -\frac{1}{r_{ij}}(\boldsymbol{r}_j \times \boldsymbol{r}_i) \cdot \boldsymbol{R}_i. \tag{6.25}$$

When $\boldsymbol{R}_i = \boldsymbol{R}_j$, it could be obtained that

$$e_{ij} \cdot \frac{\partial r_j}{\partial \omega_j} = e_{ij} \cdot \frac{\partial r_i}{\partial \omega_i}. \tag{6.26}$$

The coefficients of $\Delta\omega_i$ and $\Delta\omega_j$ in the condition equation are the same, and thus A is rank defect.

In the navigation system using inter-satellite link ranging, the coefficients of semi-major axis, eccentricity, and mean anomaly are irrelevant in most cases. When the two spacecraft are coplanar, the coefficients of inclination and argument of perigee of the two spacecraft are same, and the coefficient of the right ascension of ascending node is relevant no matter the two spacecraft are coplanar or not.

A simulation is conducted with two coplanar spacecraft in LEO orbits. The semi-major axis of the two spacecraft is 7067 km, the eccentricity is 0.003, the orbit inclination is 98.7457°, the argument of perigee is −8.9130°, the true anomaly is 0°, and the right ascension of ascending node of the two spacecraft are 74.2132° and 154.2132° respectively. Simulation results are as follows (Figs. 6.2 and 6.3).

As the simulation result shows, the navigation system just using inter-satellite link measurement can only estimate the semi-major axis, eccentricity and mean

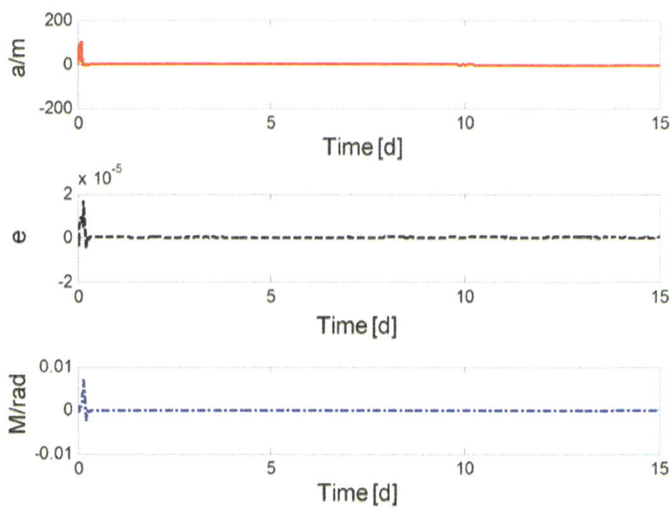

Fig. 6.2 The estimation result of a, e, M using inter-satellite link ranging

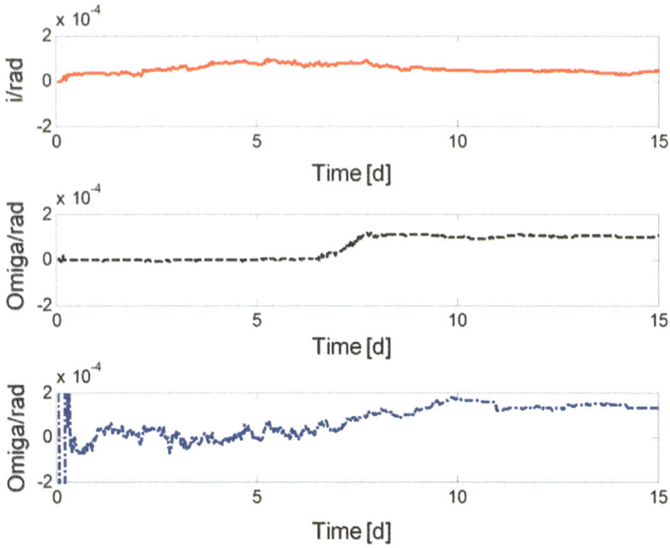

Fig. 6.3 The estimation result of i, Ω, and ω using inter-satellite link ranging

anomaly, whereas the inclination, right ascension of ascending node and the argument of perigee cannot be estimated. This is due to that the navigation system lacks an absolute external measurement, the inter-satellite link measurement remains the same when the two spacecraft perform a rotation as a whole. The inter-satellite link measurement cannot reflect the rotation of the two spacecraft, thus the inclination, right ascension of ascending node and the argument of perigee cannot be sensed. An absolute external position measurement must be added to provide the basis position information for spacecraft in order to restrain their rotation errors.

6.2 System Observation Model and Observability Analysis

6.2.1 Measurement Model for Multiple Spacecraft Observing One Pulsar

When two spacecraft receive the same pulsar signal, the time difference of arrival (TDOA) reflects the projection distance of the two spacecraft in the direction of pulsar. The observation vector can be established based upon the relative position of the two spacecraft and pulsar, as is shown in Fig. 6.4.

When spacecraft i and spacecraft j observe the same pulsar, the time of arrival (TOA) to spacecraft i and spacecraft j can be transferred to SSB respectively as follows.

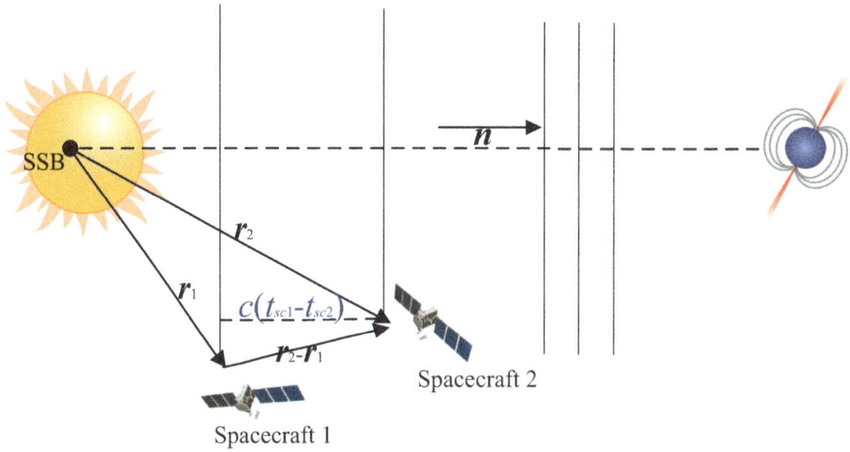

Fig. 6.4 Principle of navigation using X-ray pulsar TDOA

$$t^i_{SSB} = t^i_{SC} + \frac{1}{c}\left[\boldsymbol{n}\cdot\boldsymbol{r}^i_{SC} - \frac{\left(r^i_{SC}\right)^2}{2D_0} + \frac{\left(\boldsymbol{n}\cdot\boldsymbol{r}^i_{SC}\right)^2}{2D_0} + -\frac{\left(\boldsymbol{b}\cdot\boldsymbol{r}^i_{SC}\right)}{D_0} + \frac{(\boldsymbol{n}\cdot\boldsymbol{b})\left(\boldsymbol{n}\cdot\boldsymbol{r}^i_{SC}\right)}{D_0}\right] + \frac{2\mu_s}{c^3}\mathrm{In}\left|\frac{\boldsymbol{n}\cdot\boldsymbol{r}^i_{SC} + \boldsymbol{r}^i_{SC}}{\boldsymbol{n}\cdot\boldsymbol{b}+b} + 1\right| \tag{6.27}$$

$$t^j_{SSB} = t^j_{SC} + \frac{1}{c}\left[\boldsymbol{n}\cdot\boldsymbol{r}^j_{SC} - \frac{\left(r^j_{SC}\right)^2}{2D_0} + \frac{\left(\boldsymbol{n}\cdot\boldsymbol{r}^j_{SC}\right)^2}{2D_0} + -\frac{\left(\boldsymbol{b}\cdot\boldsymbol{r}^j_{SC}\right)}{D_0} + \frac{(\boldsymbol{n}\cdot\boldsymbol{b})\left(\boldsymbol{n}\cdot\boldsymbol{r}^j_{SC}\right)}{D_0}\right] + \frac{2\mu_s}{c^3}\mathrm{In}\left|\frac{\boldsymbol{n}\cdot\boldsymbol{r}^j_{SC} + \boldsymbol{r}^j_{SC}}{\boldsymbol{n}\cdot\boldsymbol{b}+b} + 1\right| \tag{6.28}$$

Subtracting the above two equations yields

$$\begin{aligned}\left(t^i_{SSB} - t^i_{SC}\right) - \left(t^j_{SSB} - t^j_{SC}\right) &= \frac{1}{c}\left[\boldsymbol{n}\cdot\boldsymbol{r}^{ij}_{SC} - \frac{\left(r^{ij}_{SC}\right)^2 - \left(r^{ij}_{SC}\right)^2}{2D_0} + \frac{\left(\boldsymbol{n}\cdot\boldsymbol{r}^{ij}_{SC}\right)^2 - \left(\boldsymbol{n}\cdot\boldsymbol{r}^{ij}_{SC}\right)^2}{2D_0} - \frac{\left(\boldsymbol{b}\cdot\boldsymbol{r}^{ij}_{SC}\right)}{D_0} + \frac{(\boldsymbol{n}\cdot\boldsymbol{b})\left(\boldsymbol{n}\cdot\boldsymbol{r}^{ij}_{SC}\right)}{D_0}\right] \\ &+ \frac{2\mu_s}{c^3}\mathrm{In}\left|\frac{\boldsymbol{n}\cdot\boldsymbol{r}^{ij}_{SC} + \boldsymbol{r}^{ij}_{SC} + \boldsymbol{n}\cdot\boldsymbol{b} + b}{\boldsymbol{n}\cdot\boldsymbol{r}^{ij}_{SC} + \boldsymbol{r}^{ij}_{SC} + \boldsymbol{n}\cdot\boldsymbol{b} + b}\right|\end{aligned} \tag{6.29}$$

The measurement model of two spacecraft receiving the same pulse signal can be simplified to first order, i.e.,

$$\delta t^{ij} = \frac{1}{c}\cdot n\cdot\Delta r_{ij} \tag{6.30}$$

which could be rewritten in a compact form of

$$\boldsymbol{y}_t = h(\boldsymbol{x}_t)+\boldsymbol{v}_t. \tag{6.31}$$

6.2.2 Ranging Measurement Using Inter-satellite Link

For the nth satellite in the constellation, the linearized inter-satellite link measurement model can be expressed as

$$\boldsymbol{Y}_R(t) = \boldsymbol{H}_R(\boldsymbol{X}, t) + \boldsymbol{\varepsilon}(t) \tag{6.32}$$

where

$$\boldsymbol{Y}_R(t) = \begin{cases} y_{i,1} = \rho_0^{i1} + (I^{i1} + I^{1i})/2 + \varepsilon_1(t) \\ \vdots \\ y_{i,n} = \rho_0^{in} + (I^{in} + I^{ni})/2 + \varepsilon_n(t) \end{cases} \tag{6.33}$$

where $y_{i,1}$ is the observed value of inter-satellite link between the ith satellite and the first satellite, ρ_0^{i1} is the true value of inter-satellite link between the ith satellite and the first satellite, I^{i1} is the ionospheric delay of signal from the ith satellite to the first satellite, and I^{1i} is the ionospheric delay of signal from the first satellite to the ith satellite. $\varepsilon(t)$ is the random observation noise subject to normal distribution.

And then, we have

$$\begin{aligned} \boldsymbol{H}_R^{i,j} &= \left(\frac{\partial\rho}{\partial x_i} \frac{\partial\rho}{\partial y_i} \frac{\partial\rho}{\partial z_i} 0\;0\;0 \; -\frac{\partial\rho}{\partial x_i} \; -\frac{\partial\rho}{\partial y_i} \; -\frac{\partial\rho}{\partial z_i} 0\;0\;0 \right) \\ &\quad (x_i \; y_i \; z_i \; v_{xi} \; v_{yi} \; v_{zi} \; x_j \; y_j \; z_j \; v_{xj} \; v_{xj} \; v_{xj})^T \end{aligned} \tag{6.34}$$

where $\boldsymbol{H}_R^{i,j}$ is the linearized observation matrix, which is the first order Taylor expansion of distance function at the estimated position X_0 of satellite i, and

$$\begin{cases} \frac{\partial\rho}{\partial x_{i0}} = \frac{2(x_{i0}-x_{j0})}{\rho_0} & \frac{\partial\rho}{\partial y_{i0}} = \frac{2(y_{i0}-y_{j0})}{\rho_0} & \frac{\partial\rho}{\partial z_{i0}} = \frac{2(z_{i0}-z_{j0})}{\rho_0} \\ \frac{\partial\rho}{\partial x_{j0}} = -\frac{2(x_{i0}-x_{j0})}{\rho_0} & \frac{\partial\rho}{\partial y_{j0}} = -\frac{2(y_{i0}-y_{j0})}{\rho_0} & \frac{\partial\rho}{\partial z_{j0}} = -\frac{2(z_{i0}-z_{j0})}{\rho_0} \end{cases}. \tag{6.35}$$

6.2.3 Observability Analysis

The system observability indicates whether the system state can be determined by system measurement. In this section, we exploit the system observability by analyzing the system measurement model which could reflect the connection between the measurement and the system state. If the partially differential measurement model w.r.t one component of the system state equals 0, the corresponding system state is viewed to be not observable.

As shown in Sect. 6.1.2, the inter-satellite link measurement is not observable for the right ascension of ascending node, coplanar orbit inclination and argument of perigee. We will show those orbital elements could be sensed by the pulsar-based observation.

(1) **Right ascension of ascending node** Ω

The partially differential equation of pulsar-based observation model with respect to Ω is expressed as

$$\frac{\partial \boldsymbol{n} \cdot \Delta \boldsymbol{r}}{\partial \Omega} = \frac{\partial \boldsymbol{n}_x \Delta x}{\partial \Omega} + \frac{\partial \boldsymbol{n}_y \Delta y}{\partial \Omega} + \frac{\partial \boldsymbol{n}_z \Delta z}{\partial \Omega} \tag{6.36}$$

$$\frac{\partial \boldsymbol{n} \cdot \Delta \boldsymbol{r}}{\partial \Omega} = \boldsymbol{n}_x \frac{\partial x_i}{\partial \Omega} + \boldsymbol{n}_y \frac{\partial y_i}{\partial \Omega} + \boldsymbol{n}_z \frac{\partial z_i}{\partial \Omega} - (\boldsymbol{n}_x \frac{\partial x_j}{\partial \Omega} + \boldsymbol{n}_y \frac{\partial y_j}{\partial \Omega} + \boldsymbol{n}_z \frac{\partial z_j}{\partial \Omega}). \tag{6.37}$$

Given $\frac{\partial \boldsymbol{r}}{\partial \Omega} = \boldsymbol{J}_z \times \boldsymbol{r} = \begin{pmatrix} -y \\ x \\ 0 \end{pmatrix}$, we have

$$\frac{\partial \boldsymbol{n} \cdot \Delta \boldsymbol{r}}{\partial \Omega} = -\boldsymbol{n}_x y_i + \boldsymbol{n}_y x_i - (-\boldsymbol{n}_x y_j + \boldsymbol{n}_y x_j) = \boldsymbol{n}_y (x_i - x_j) - \boldsymbol{n}_x (y_i - y_j) \tag{6.38}$$

where $\boldsymbol{n}$ is the direction vector of pulsar, $\Delta \boldsymbol{r}$ is the relative position between two spacecraft, Δx, Δy and Δz are the three components of the relative position in the inertial system, $\boldsymbol{n}_x$, $\boldsymbol{n}_y$ and $\boldsymbol{n}_z$ are the directions of those components, $\boldsymbol{r}_i$, $\boldsymbol{r}_j$ are the positions of spacecraft, and x_i, y_i, z_i, x_j, y_j and z_j are the components.

According to Eq. (6.38), when $\frac{(x_i - x_j)}{(y_i - y_j)} \neq \frac{\boldsymbol{n}_x}{\boldsymbol{n}_y}$, the partially difference of pulsar-based measurement model with respect to Ω is not equal to 0, so the measurement model is observable for it.

(2) **Inclination** i

The partially differential equation of pulsar-based measurement model with respect to i is expressed as

$$\frac{\partial \boldsymbol{n} \cdot \Delta \boldsymbol{r}}{\partial i} = \frac{\partial \boldsymbol{n}_x \Delta x}{\partial i} + \frac{\partial \boldsymbol{n}_y \Delta y}{\partial i} + \frac{\partial \boldsymbol{n}_z \Delta z}{\partial i} \tag{6.39}$$

$$\frac{\partial \boldsymbol{n}\cdot\Delta\boldsymbol{r}}{\partial i}=\boldsymbol{n}_x\frac{\partial x_i}{\partial i}+\boldsymbol{n}_y\frac{\partial y_i}{\partial i}+\boldsymbol{n}_z\frac{\partial z_i}{\partial i}-(\boldsymbol{n}_x\frac{\partial x_j}{\partial i}+\boldsymbol{n}_y\frac{\partial y_j}{\partial i}+\boldsymbol{n}_z\frac{\partial z_j}{\partial i}). \tag{6.40}$$

Given $\frac{\partial \mathbf{r}}{\partial i}=\boldsymbol{J}_N\times\boldsymbol{r}=\begin{pmatrix}\cos\Omega\\ \sin\Omega\\ 0\end{pmatrix}\times\boldsymbol{r}$, we have

$$\frac{\partial \boldsymbol{n}\cdot\Delta\boldsymbol{r}}{\partial i}=\boldsymbol{n}\cdot(\boldsymbol{J}_{Ni}\times\boldsymbol{r}_i)-\boldsymbol{n}\cdot(\boldsymbol{J}_{Nj}\times\boldsymbol{r}_j)=(\boldsymbol{n}\times\boldsymbol{r}_i)\cdot\boldsymbol{J}_{Ni}-(\boldsymbol{n}\times\boldsymbol{r}_j)\cdot\boldsymbol{J}_{Nj}. \tag{6.41}$$

If $\Omega_i=\Omega_j$, then $\boldsymbol{J}_{Ni}=\boldsymbol{J}_{Nj}$,

$$\begin{aligned}\frac{\partial(\mathbf{n}\cdot\Delta\mathbf{r})}{\partial i}&=(\mathbf{n}\times\mathbf{r}_i)\cdot\mathbf{J}_N-(\mathbf{n}\times\mathbf{r}_j)\cdot\mathbf{J}_N\\ &=\mathbf{J}_N\cdot((\mathbf{n}\times\mathbf{r}_i)-(\mathbf{n}\times\mathbf{r}_j))=\mathbf{J}_N\cdot(\mathbf{n}\times(\mathbf{r}_i-\mathbf{r}_j))\\ &=(\mathbf{n}_y\Delta\mathbf{r}_z-\mathbf{n}_z\Delta\mathbf{r}_y)\cos\Omega+(\mathbf{n}_z\Delta\mathbf{r}_x-\mathbf{n}_x\Delta\mathbf{r}_z)\sin\Omega\end{aligned} \tag{6.42}$$

According Eq. (6.42), when $\frac{(\boldsymbol{n}_y\Delta\boldsymbol{r}_z-\boldsymbol{n}_z\Delta\boldsymbol{r}_y)}{(\boldsymbol{n}_z\Delta\boldsymbol{r}_x-\boldsymbol{n}_x\Delta\boldsymbol{r}_z)}\neq\frac{\sin\Omega}{\cos\Omega}$, the partially difference of pulsar-based measurement model with respect to i is not equal to 0, so the measurement model is observable for it.

(3) **Argument of Perigee,** ω

The partially differential equation of pulsar-based measurement model with respect to ω is expressed as

$$\frac{\partial \boldsymbol{n}\cdot\Delta\boldsymbol{r}}{\partial \omega}=\boldsymbol{n}\cdot(\boldsymbol{R}_i\times\boldsymbol{r}_i)-\boldsymbol{n}\cdot(\boldsymbol{R}_j\times\boldsymbol{r}_j)=(\boldsymbol{n}\times\boldsymbol{r}_i)\cdot\boldsymbol{R}_i-(\boldsymbol{n}\times\boldsymbol{r}_j)\cdot\boldsymbol{R}_j \tag{6.43}$$

where $\boldsymbol{R}$ is the orbital normal unit vector with expression of $\boldsymbol{R}=\frac{1}{\sqrt{\mu p}}(\boldsymbol{r}\times\dot{\boldsymbol{r}})$.

If $\boldsymbol{R}_i=\boldsymbol{R}_j$, then

$$\frac{\partial \boldsymbol{n}\cdot\Delta\boldsymbol{r}}{\partial \omega}=(\boldsymbol{r}_i\times\boldsymbol{n})\cdot\boldsymbol{R}-(\boldsymbol{r}_j\times\boldsymbol{n})\cdot\boldsymbol{R}=\boldsymbol{r}_i\cdot(\boldsymbol{n}\times\boldsymbol{R})-\boldsymbol{r}_j\cdot(\boldsymbol{n}\times\boldsymbol{R})=\Delta\boldsymbol{r}_{ij}\cdot(\boldsymbol{n}\times\boldsymbol{R}) \tag{6.44}$$

When $\Delta\boldsymbol{r}_{ij}\cdot(\boldsymbol{n}\times\boldsymbol{R})\neq 0$, $\frac{\partial\boldsymbol{n}\cdot\Delta\boldsymbol{r}}{\partial\omega}\neq 0$, and the pulsar-based measurement model is observable for ω.

6.3 Satellite Constellation Autonomous Navigation Using TDOA of Pulsar

6.3.1 Scheme Design

Assume two satellites A and B in the constellation observe one pulsar at the same time. After a certain period of observation, the TDOA of the pulsar signal at two satellites can be obtained. The TDOA multiplying the speed of light could reflect the distance between the two satellites along the direction of pulsar. Given that the distance between A and B can be measured by inter-satellite link, the angle θ between the direction of pulsar and the baseline AB, as shown in Fig. 6.5, can be calculated. Regarding that θ reflects the angle between the baseline and the inertial system, the absolute direction of the whole satellite constellation could be estimated if θ is calculated.

We provide a scheme to determine the constellation orientation parameters using single baseline with single TDOA observation.

Figure 6.6 illustrates the observation scheme. Two satellites (e.g. satellite 1 and satellite 2) in the constellation form a baseline. The two satellites load one X-ray sensor, respectively. Adjust the X-ray sensors on the two satellites to ensure they could point at one common pulsars at the same time. Calculate the direction of such baseline vector in the inertial space by using the TDOA measurement. And then, estimate the positions of all the satellites in the constellation by using the inter-satellite link measurement and TDOA. The scheme flow is as shown in Fig. 6.7.

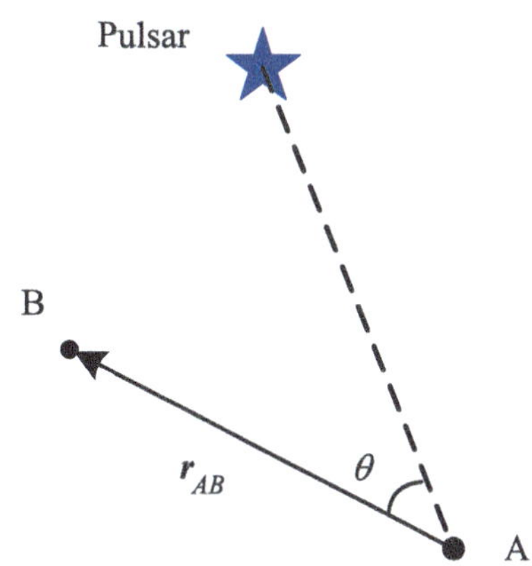

Fig. 6.5 Diagram of inter-satellite baseline vector and pulsar direction

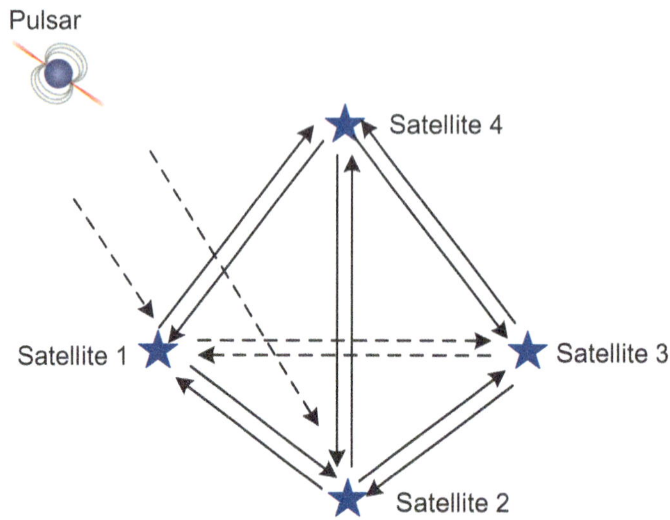

Fig. 6.6 Diagram of synchronized observation of the same pulsar

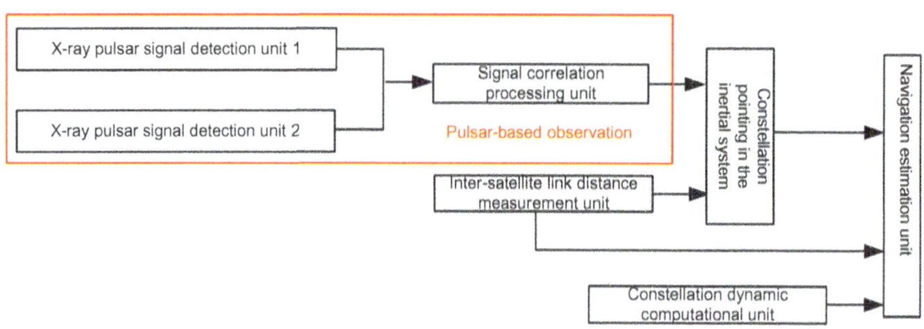

Fig. 6.7 Scheme flow chart (single direction observation)

Table 6.1 Initial orbital elements of constellation (single direction observation)

No.	Orbit type	Semi-major axis (km)	Eccentricity	Inclination (°)	Right ascension of ascending node (°)	True anomaly (°)
1	MEO	22,000	0	55	0	20
2	MEO	22,000	0	55	0	135
3	MEO	22,000	0	55	0	270
4	MEO	22,000	0	55	120	45
5	MEO	22,000	0	55	120	180
6	MEO	22,000	0	55	120	305
7	MEO	22,000	0	55	240	90
8	MEO	22,000	0	55	240	360

6.3.2 Simulation Analysis

The initial orbital elements of satellite are shown in Table 6.1. The dynamic model error is 10^{-7} m/s^2, the distance measurement error is 0.1 m, the filter is the federated type, the X-ray pulsar used in navigation is PSR B0531+21, and the simulation duration is 20 days.

As can be seen from Figs. 6.8 to 6.9, the estimations of i, Ω are convergent. The mean variance of Ω and i are 9.9 mas and 26 mas, respectively.

Under the above satellite constellation, adding the one pulsar PSR B1957+20 for observation and keeping other conditions unchanged, Figs. 6.10 and 6.11 show the results.

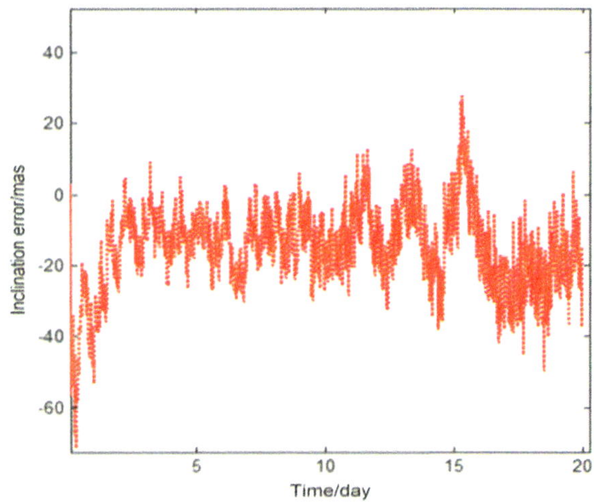

Fig. 6.8 Inclination error

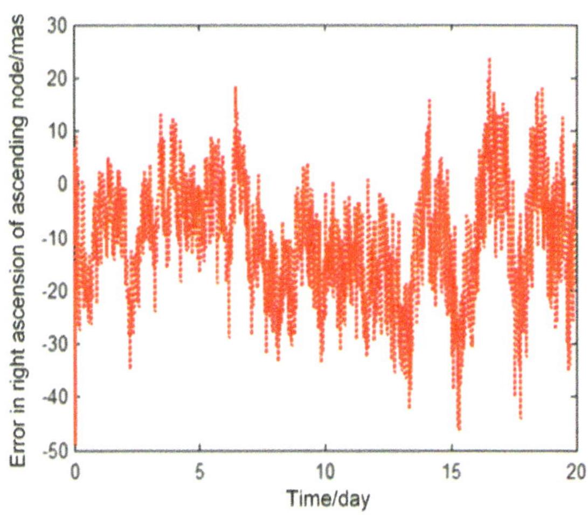

Fig. 6.9 Error in right ascension of ascending node of orbit

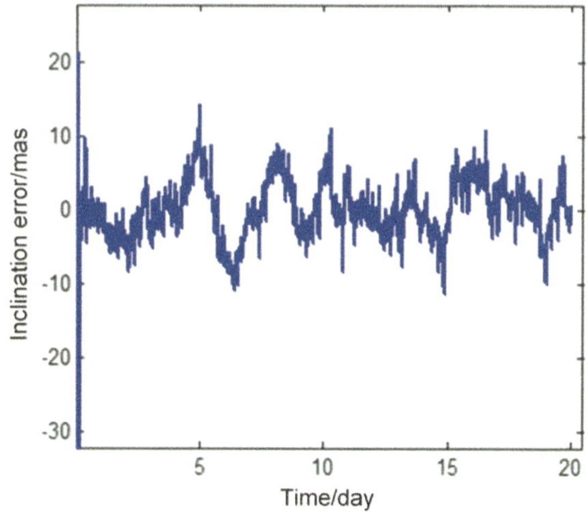

Fig. 6.10 Inclination error

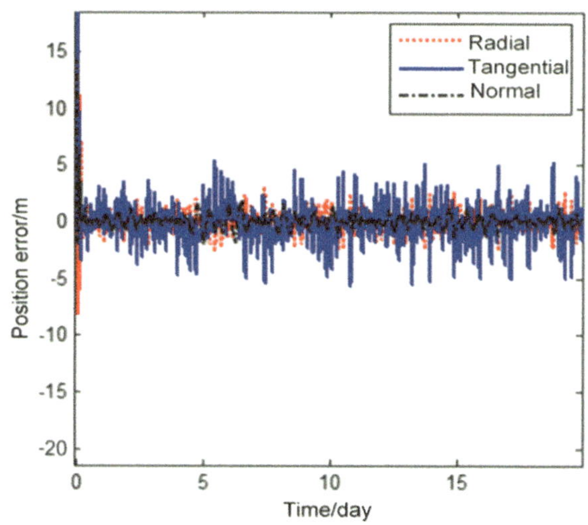

Fig. 6.11 Navigation accuracy

As can be seen from Figs. 6.10 and 6.11, the estimations of i, Ω remain convergent. The mean variance of Ω and i are 5 mas and 16 mas, respectively.

6.4 Spacecraft Autonomous Navigation Network

An interplanetary autonomous navigation network is proposed in this section, which is named as the Internet of Spacecraft (IoS) and enables a solar-system wide autonomous navigation capability for spacecraft. This network consists of two types of spacecraft, namely, host spacecraft and client spacecraft. The former provide the absolute reference for the whole network, and the latter determine their positions by communicating with the host spacecraft or other client spacecraft. To investigate the performance of IoS, a detailed application scheme that supports the flight from Earth to Mars is developed and analyzed.

6.4.1 Framework of IoS

IoS is an interplanetary autonomous navigation network which can cover the whole solar system. Spacecraft of IoS can be divided into two parts, namely, host spacecraft and client spacecraft. The host spacecraft are settled nearby some planets such as the Earth, Mars and so on. They form the absolute reference of IoS and their positions can be determined by the method mention in Sect. 6.3. The client spacecraft are launched for some space missions on which the intersatellite ranging equipment are installed to communicate with others and they are positioned by intersatellite links. There are three navigation and communication modes for client spacecraft, which can be programmed for different flight phases. These three modes are described as follows.

Mode 1: the client spacecraft determine their own positions by communicating with the host spacecraft.
Mode 2: the client spacecraft obtain navigation information from the host spacecraft as well as the client spacecraft.
Mode 3: the client spacecraft only communicate with other client spacecraft to get navigation information.

Moreover, this network is not a simple extension of the navigation constellation. It possesses several characteristics and advantages. Firstly, the reference of IoS is independent of the ground-based tracking and control system. Their positions and velocities can be got by the autonomous navigation method using TDOA of pulsars. Secondly, configuration of this network and numbers of the client spacecraft are flexible. We do not need to design the configuration of the whole network, and the spacecraft installed with compatible intersatellite ranging equipment can be added in this network and obtain autonomous navigation service by communication with others. Finally, arbitrary spacecraft of IoS can provide navigation information for the others. After determining their own positions, spacecraft of IoS can transmit their position information to the others. Therefore, these client spacecraft can also be treated as host spacecraft.

Figure 6.12 shows the concept sketch of IoS that can provide navigation information in the whole solar system. In future, the amount of DSEs or interplanetary spacecraft will be continuously increased. While they implement their own tasks, they can also be treated as newly joined spacecraft of IoS at the same time, thus there would be more and more spacecraft in the IoS. As a result, this network would be upgraded

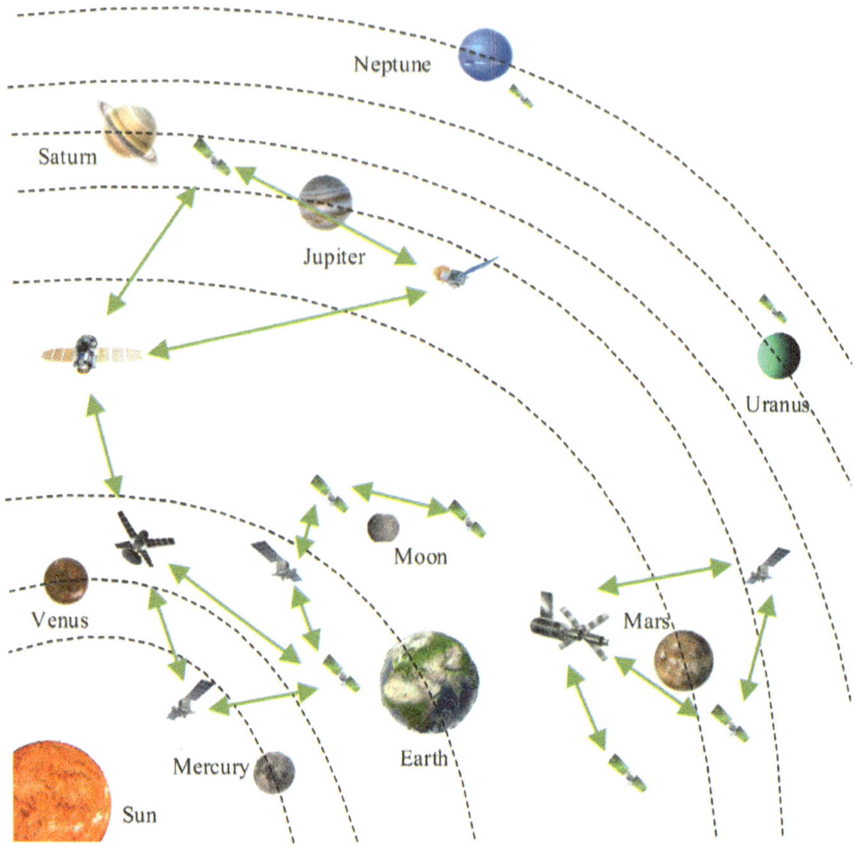

Fig. 6.12 The concept sketch of IoS that covers the whole solar system

and its coverage could be expanded to further places. Finally, a highly reliable, efficient and low-cost interplanetary autonomous navigation network is expected to be formed.

6.4.2 *A Detailed Design for IoS that Support the Flight from the Earth to Mars*

6.4.2.1 Design of the Reference of IoS

According to the analysis of Sect. 6.4.1, host spacecraft are absolute references of the whole network. To design the reference of IoS, the space region of client spacecraft should be ascertained firstly. Then, we need to design the orbit of the reference to realize geometrically coverage of client spacecraft. Next, the physical coverage should be guaranteed, that is, the signal intensity should be strong enough to be received. Finally, high-accuracy autonomous navigation methods should be applied for host spacecraft, for which the autonomous navigation method using TDOA of pulsars can be applied. Therefore, we design the reference of IoS through these two aspects: (1) design of the orbit, and (2) design of the physical coverage.

(1) Design of the orbit

The space region of client spacecraft, geometrical configuration and orbital stability are taken into account to design the orbit.

Firstly, considering the target users are Mars probes. Their mission phases can usually be divided into the launching phase, cruising phase, and round phase [1]. As shown in Fig. 6.13 the launching phase is located in the Earth's sphere of influence. Then the spacecraft takes a long journey along an interplanetary orbit around sun, which is the cruising phase. After that, the spacecraft inserts into the orbit around Mars. Obviously, all of these phases should be covered by IoS.

Then, considering that host spacecraft should be placed nearby the planet and it is influenced by the gravitational force of the sun, thus the dynamical model of host spacecraft can be regarded as the circular restricted three body problem (CRTBP). The libration points of this model can be used to deploy the space telescope and station

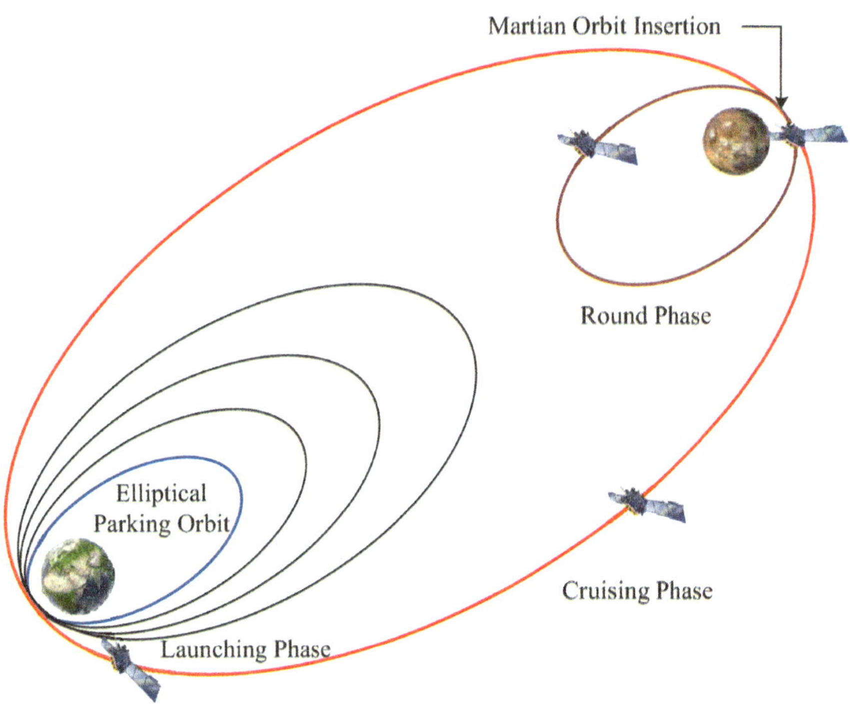

Fig. 6.13 Schematic of flight phases in Mars exploration

for future lunar or deep space exploration missions. Moreover, the communication and navigation constellation can be arranged around these points to support the autonomous navigation for DSEs [2, 3]. So the Sun-Earth libration points and the Sun-Mars libration points are chosen for the host spacecraft.

Figure 6.14 shows the relative location of Sun-Earth libration points and Sun-Mars libration points, which are respectively denoted as EL_1–EL_5 and ML_1–ML_5 [4]. Considering the long distances from the libration points L_3, L_4 and L_5 to their main body, the signal propagation loss will be enormous. In addition, the transfer trajectories of Mars probes are mostly direct trajectories [5], and with consideration of energy, the electromagnetic beam angle should be limited, thus the host spacecraft should be put at only one side of Earth or Mars.

Finally, considering that the Halo orbit around the libration points is not sheltered form the celestial objects and its amplitude in the normal direction of the primary body's revolution orbits is large enough. Therefore, the host spacecraft located at this

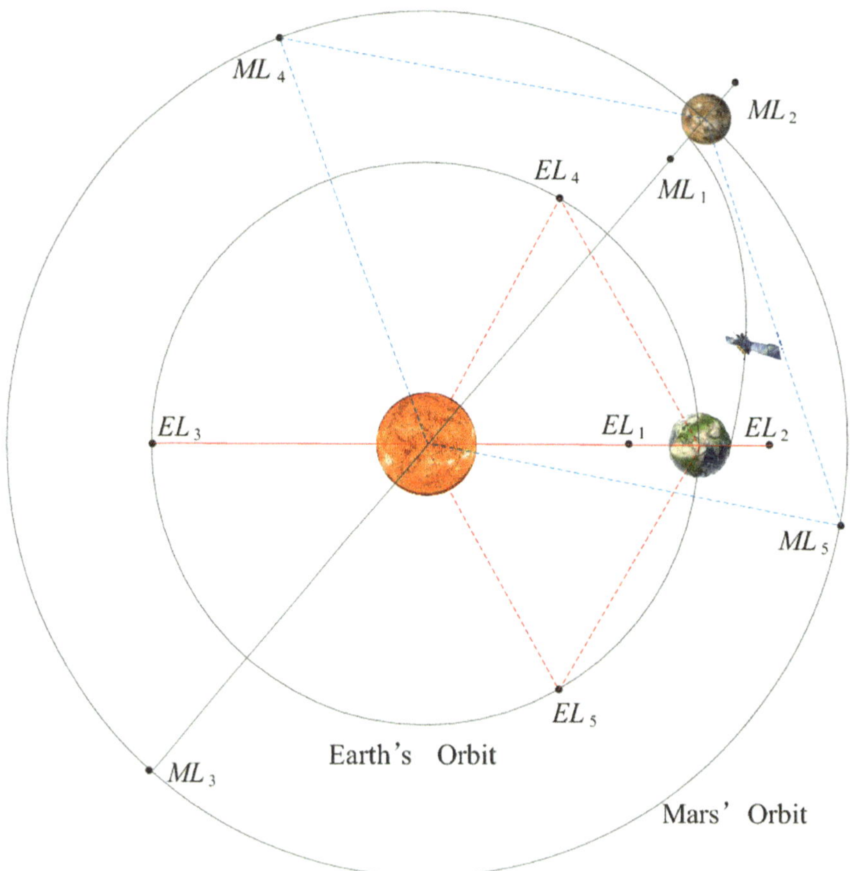

Fig. 6.14 Libration points of Sun-Earth and Sun-Mars systems

orbit possess higher dispersity and smaller linearity. Moreover, the stability of this orbit is better. Therefore, the Halo orbits around EL_2 (or EL_1) and ML_2 (or ML_1) are chosen to put host spacecraft. Figure 6.14 shows the specific space configuration scheme of host spacecraft by taking the L_2 libration points as an example.

(2) Design of the reference's physical coverage

Because of the broad geometrical coverage and far operating distance of IoS, the problem of signal propagation is very serious. And it is necessary to analyze the intensity of the navigation signal to ensure the physical coverage.

Generally speaking, navigation signals propagate in the form of electromagnetic waves, and formula of the signal propagation loss is

$$Los = 32.44 + 20\lg d(\text{km}) + 20\lg f(\text{MHz}) \tag{6.45}$$

where f is the operating frequency and d is the propagation distance. When f or d is doubled, the signal propagation loss will increase by 6 dB. So the maximum signal loss can be analyzed to reflect the effective navigation distance of the reference.

GPS is the mostly used global navigation system and satellites of GPS can be analyzed to help confirm the corresponding parameters of host spacecraft. Table 6.2 shows some parameters of the satellite of GPS [6]. Considering the development of the science and technology, the corresponding parameters could be improved and the navigation distance would be consequently lengthened.

The intensity of the emission signal is related to the beam angle. And the formula of the signal gain relative to the beam angle α is $10 \times \lg G_T(\alpha)$, where

$$G_T(\alpha) = \frac{2}{1 - \cos\alpha}. \tag{6.46}$$

Because of the long operation distance of the host spacecraft, its beam angle does not need to be so large as that of GPS, and a 6-degree angle is enough to cover the range of 78500 km from the geocenter and 54900 km from the Mars center. According to Eq. (6.46), the signal intensity can be strengthened to 25.6 dB, which is improved by 11 dB compared to GPS. In addition, in order to enhance the anti-interference ability, the transmitting power of GPS-III satellites will improve by one hundred times [7], and the input power of the satellite antenna will consequently improve by 20 dB. Moreover, the SNR of the receiver may reach 25 dB/Hz by applying the weak signal acquisition technique [8, 9].

Table 6.2 Parameters of the GPS satellites

Parameters	Value
Beam angle	21.3°
Transmit antenna gain	14.7 dB
Signal to noise ratio (SNR) of received signal	40 dB/Hz
Navigation distance	20180 km

On the basis of the above analysis, the propagation path loss can increase 46 dB than that of GPS, and the navigation distance is about 200 times farther than GPS, that is, 4×10^6 km, which is much longer than the distance from EL_2 and ML_2 to Earth and Mars respectively (Fig. 6.15).

Then we utilize the simulated data to analyze whether the effective navigation distance can cover the cruising phase. Table 6.3 shows the initial state of a transfer trajectory from the Earth to Mars in the heliocentric inertial coordinate system [5]. Combining the planetary ephemeris DE405, this trajectory is simulated and it is shown in Fig. 6.16.

Figure 6.17 shows the distance from this transfer trajectory to EL_2 and ML_2. As shown in Fig. 6.17, the configuration scheme shown in Fig. 6.15 can provide navigation information for the initial and end of the cruising phase, which is significant that the Mars probe can get accurate insertion state.

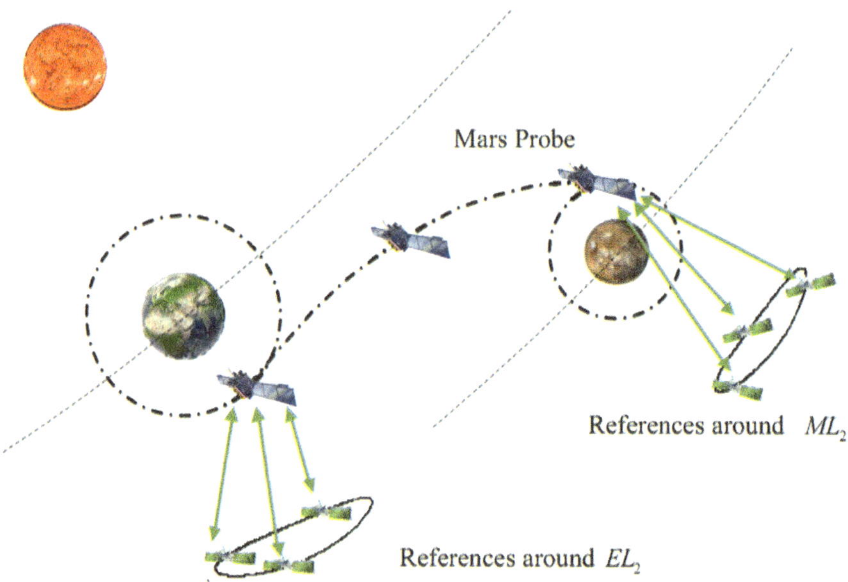

Fig. 6.15 The configuration of the host spacecraft around EL_2 and ML_2

Table 6.3 Initial states of an Earth-Mars transfer orbit in the heliocentric inertial coordinate

State-vector component	Value
X [km]	4.4825×10^7
Y [km]	1.2887×10^8
Z [km]	5.5869×10^7
$\dot{X}$ [km/s]	−31.5034
$\dot{Y}$ [km/s]	7.2765
$\dot{Z}$ [km/s]	4.7646

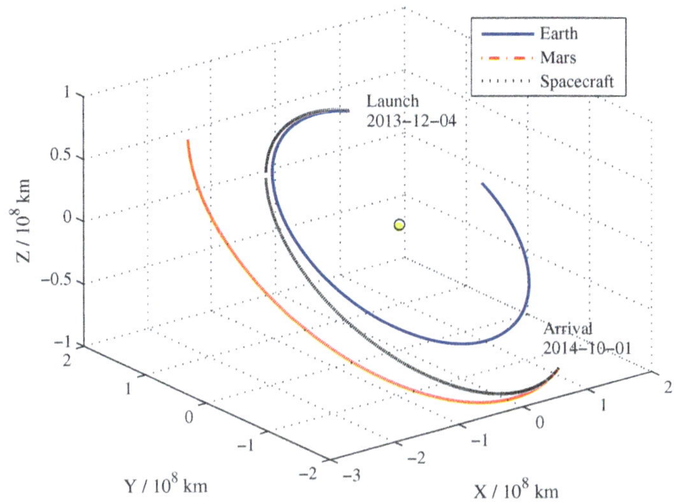

Fig. 6.16 The transfer trajectory from the Earth to Mars in the heliocentric inertial coordinate system

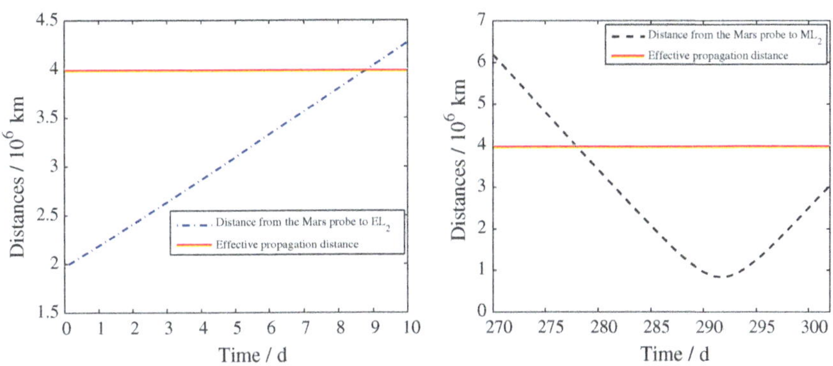

The first 10 days of the cruising phase   The last 34 days of the cruising phase

Fig. 6.17 Distances from the Mars probe to EL_2 and ML_2

6.4.2.2 Navigation Modes of Client Spacecraft

After the host spacecraft determines their own positions, they can provide navigation service for the client spacecraft, which only needs to carry the intersatellite ranging equipment. Moreover, note that when one client spacecraft has been positioned, its position can be transmitted to others in the form of directional through the "request-response" mode.

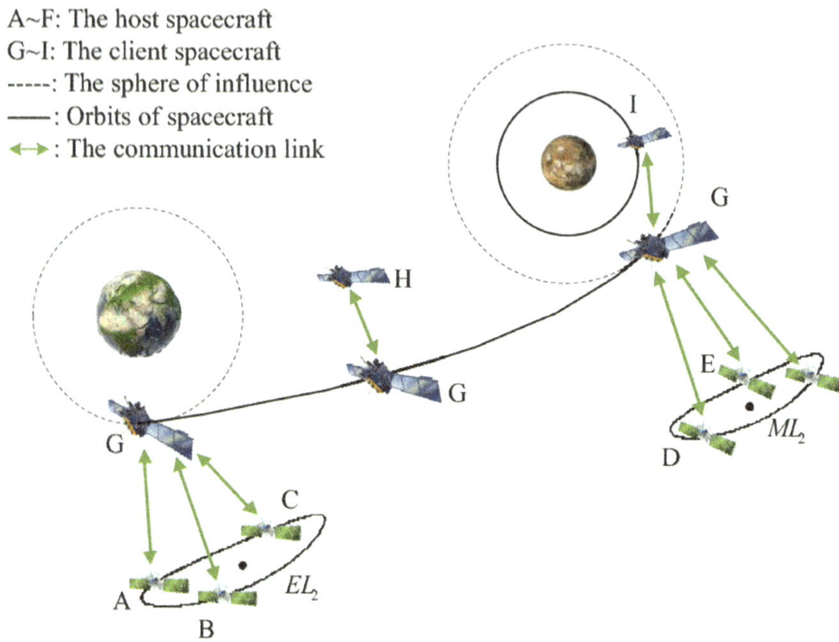

Fig. 6.18 Schematic of the client spacecraft's navigation modes in the IoS system that support the flight from the Earth to Mars

With the increase of space missions, IoS will be continuously expanded and improved. The subsequent spacecraft can program the communication links according to their own requirements and choose one or several nearby spacecraft of IoS to realize its own autonomous navigation.

Combining the designed scheme of the reference of IoS in Sect. 6.4.2, navigation modes of the client spacecraft are analyzed. Figure 6.17 shows different navigation modes of the Mars probe at different phase, which have been mentioned in Sect. 6.4.1. A, B and C are host spacecraft at the Halo orbit around EL_2, D, E and F are the reference around ML_2, the spacecraft G is the Mars probe whose communication links has been programmed, H is another client spacecraft that within the communication scope of G, and I is a Mars Orbiter that can provide navigation information to others.

The Mars probe G can choose different host spacecraft or client spacecraft to communicate with at different flight phases and three different cases are designed to investigate the performance of the corresponding navigation modes.

Case 1: The spacecraft G is located at the initial stage of the cruising phase, and can be covered by the host spacecraft. Therefore, it can choose the host spacecraft around EL_2 as the communication target.
Case 2: The spacecraft G's flight segment is still the cruising phase, but it cannot receive the signal of host spacecraft in this scheme. Then it could choose other client

spacecraft (the spacecraft H in Fig. 6.18) within its communication scope to modify its state.

Case 3: The client spacecraft G has inserted into Mars' sphere of influence and the host spacecraft around ML_2 and a Mars Orbiter (the spacecraft I in Fig. 6.18) are chosen to provide navigation service.

6.4.3 *Simulation Analysis*

In this section, the navigation results of the reference by using TDOA of pulsars and the investigation of three cases mentioned in Sect. 6.4.2.2 are simulated.

6.4.3.1 Simulation for the Navigation Method Applied to the Reference of IoS

The autonomous navigation method using TDOA of pulsar is used for the reference of IoS. Table 6.4 shows the parameters of the chosen pulsars [10], and the x-ray background radiation flux is taken as 0.005 ph/cm^2/s [11]. Table 6.5 shows the initial state of the Halo orbits around EL_2 and ML_2 in the synodic reference frame [12], where [L] and [T] are respectively the normalized unit of length and time. The z-axis amplitude of the Halo orbit is 2×10^5 km, and host spacecraft are distributed every equal phase. The initial state errors are respectively taken as 10 km and 10 m/s, the error of intersatellite links is 1 m, and the observation period is 3600 s.

Then, the Unscented Kalman Filter (UKF) is applied to investigate the performance of this navigation method. Figure 6.19 shows the position estimated errors and 3 sigma outlines of the host spacecraft around EL_2 and ML_2, and it can be seen that the positional accuracy can reach the order of 50 m.

Table 6.4 Parameters of chosen pulsars

Parameters	Pulsars		
	B0531+21	B1821−24	B1937+21
Right ascension angle [deg]	83.63	276.13	294.92
Declination angle [deg]	22.01	−24.87	21.58
P [s]	0.0334	0.00305	0.00156
W [s]	1.7×10^{-3}	5.5×10^{-5}	2.1×10^{-5}
F_x [ph/cm^2/s]	1.54	1.93×10^{-4}	4.99×10^{-5}
p_f [%]	70	98	86

Table 6.5 Initial states of the Halo orbit around EL_2 and ML_2

State parameters	Location of the Halo orbit	
	Around EL_2	Around ML_2
x_0/[L]	1.008320655353143	1.003857250806226
y_0/[L]	0	0
z_0/[L]	0.000744256879882	0.000737122161894
v_{x0}/[L/T]	0	0
v_{y0}/[L/T]	0.009958906346744	0.005044116911458
v_{z0}/[L/T]	0	0

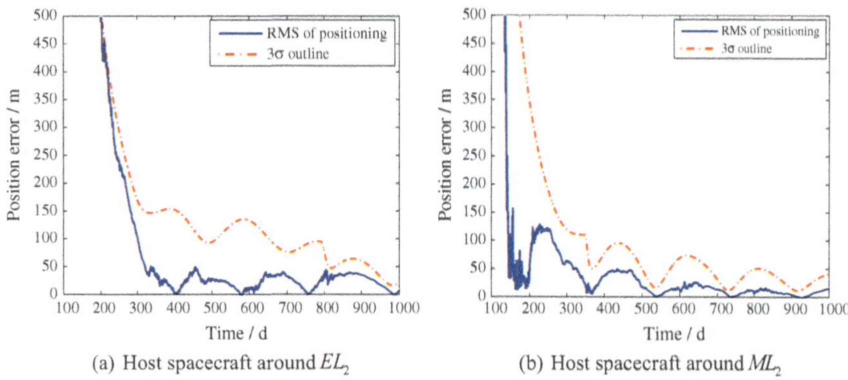

Fig. 6.19 The position estimation error of host spacecraft around EL_2 and ML_2

6.4.3.2 Simulation for the Navigation Modes of Client Spacecraft

Three different cases mentioned in Sect. 6.4.2.2 are simulated in this sub section.

For case 1, the initial states of spacecraft G (shown in Fig. 6.19) are the same as that in Table 6.2, and its position and velocity errors are respectively 10 km and 10 m/s. The navigation interval is 3600 s, and the error of intersatellite links is taken as 50 m considering the position error of host spacecraft which has been analyzed in Sect. 6.4.2.1. Figure 6.20 shows the result of UKF and the 3 sigma outlines. It can be seen that the position error can converged within 6 days and reach the precision of about 1 km. Therefore, the Mars probe can insert the cruising phase with smaller state errors.

Table 6.6 shows the initial states of spacecraft G and H for case 2. The simulation conditions are taken as the same as that in case 1. And the positioning result is shown in Fig. 6.21, which can reach an accuracy of 40 km (3σ).

Table 6.7 shows the orbit elements of the Mars probe G and the Orbiter I, the position errors and velocity errors are taken as 1 km and 1 m/s. And the other

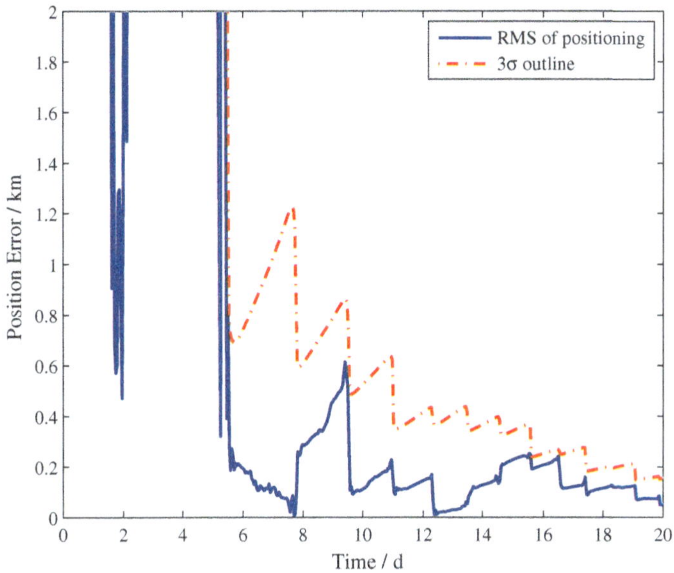

Fig. 6.20 The position error of the client spacecraft for case 1

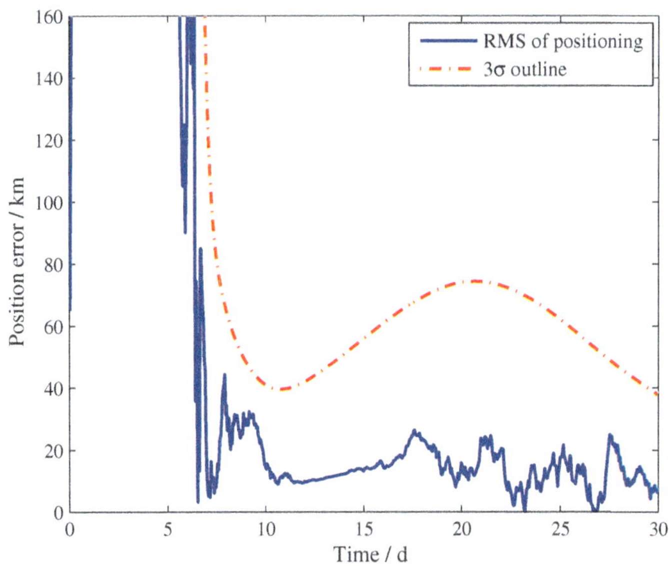

Fig. 6.21 The positioning error of the client spacecraft for case 2

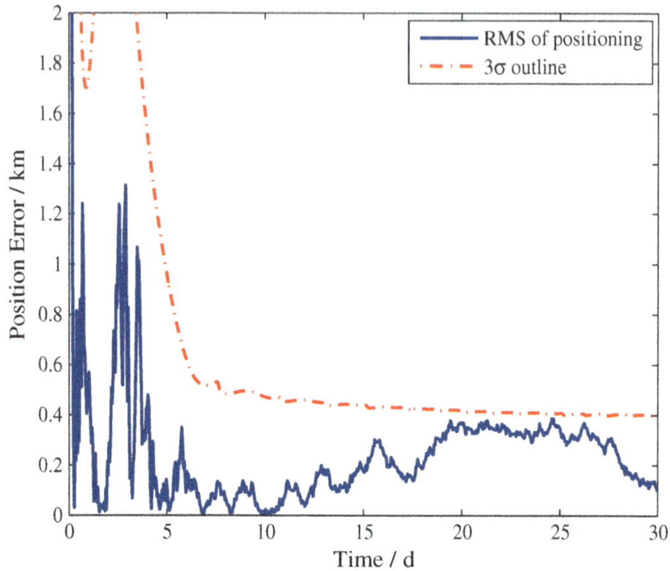

Fig. 6.22 The position error of the client spacecraft for case 3

Table 6.6 Initial states of the spacecraft G and H in the heliocentric inertial coordinate system

State parameters	G	H
X [km]	5.8903×10^7	5.8845×10^7
Y [km]	1.2503×10^8	1.2505×10^8
Z [km]	5.3475×10^7	5.7786×10^7
$\dot{X}$ [km/s]	−30.5362	−30.5568
$\dot{Y}$ [km/s]	9.6449	9.5980
$\dot{Z}$ [km/s]	5.7845	−3.7258

Table 6.7 Orbit elements of the client spacecraft that are near Mars [13]

Orbit elements	G	I
Semi-major axis [km]	−5432.0	3684.5
Eccentricity	1.617	0.010
Inclination [deg]	44.1	93.0
Longitude of node [deg]	91.9	278.0
Argument of periapsis [deg]	297.2	270.0

simulation conditions are the same as that in case 1. Figure 6.22 shows the results of this case, and the position errors are within 500 m.

The above three cases represent three different navigation and communication modes of the client spacecraft of IoS. Results show that the IoS can provide high-accuracy navigation service for the Mars probe during its whole flight. Furthermore, this continuously updated network can play an important role for the mission to Mars as well as other deep space exploration missions.

6.5 Summary

Taking account of the shortcomings of inter-satellite link, the autonomous navigation method using TDOA of pulsar is described in this chapter. Firstly, its principle is given and secondly its observability is analyzed. Then, this navigation method is used for the satellite constellation. Finally, an interplanetary autonomous navigation network named IoS is proposed and it is consisted of client spacecraft and host spacecraft. The autonomous navigation method using TDOA of pulsar can be applied for the client spacecraft, so that they can provide high-accuracy and reliable navigation information for the whole network.

References

1. Miele A, Wang T (1999) Optimal transfers from an Earth orbit to a Mars orbit. Acta Astronaut 45(3):119–133
2. Bhasin K, Hayden J (2004) Developing architectures and technologies for an evolvable NASA space communicationin frastructure. 22th AIAA International communications satellite system conference and exhibit, Monterey, pp 1180–1191
3. Kulkarni TR, Dharne A, Mortari D (2005) Communication architecture and technologies for missions to moon, mars, and beyond. Space exploration conference: continuing the voyage of discovery, Orlando, Florida, AIAA, pp 2005–2778
4. Meng YH, Zhang YD, Chen QF (2015) Dynamics and control of spacecraft near libration points. Science Press, Beijing
5. Luo HJ (2013) Orbit design and optimization of Mars probe. MSc thesis. Harbin Institute of Technology, Harbin
6. Meng YH, Chen QF (2014) Outline design and performance analysis of navigation constellation near earth-moon libration point. Acta Physica Sinica 63(24):1429–1438
7. Sun HT, Wang CQ, Feng JD (2004) Current situation and development of anti-jamming technology of GPS system in USA. Electro-Optic Technol Appl 19(3):57–63
8. Grant H, Dodds D (2009) A new comparison of averaging techniques used for weak signal acquisition with application to GPS L5 signals. In: Proceedings of the Institute of Navigation GNSS conference, Savannah, GA, pp 2610–2616
9. Karunanayake MD, Cannon ME, Lachapelle G (2004) Evaluation of assisted GPS (AGPS) in weak signal environments using a hardware simulator. In: Proceedings of the Institute of Navigation GNSS conference, Long Beach, California, pp 2416–2426
10. Sheikh SI, Pines DJ (2006) Recursive estimation of spacecraft position and velocity using x-ray pulsar time of arrival measurements. J Inst Navig 53:149–166

11. Dennis WW (2005) The use of x-ray pulsars for aiding GPS satellite orbit determination. Master dissertation, Air Force Institute of Technology
12. Parker J, Anderson R, Born G, Fujimoto K, Leonard J, McGranaghan R (2006) Navigation between geosynchronous and lunar L1 Orbiters. AIAA/AAS astrodynamics specialist conference, pp 13–16
13. Lightsey E, Mogensen A, Burkhart P et al (2008). Real-time navigation for Mars missions using the Mars network. J Spacecraft Rockets 45(3):519–533

Chapter 7
Ground-Based Simulation and Verification System for X-Ray Pulsar-Based Navigation

As the X-ray pulsar signal cannot travel through the atmosphere around the Earth, it is difficult to investigate the performance of X-ray pulsar-based navigation system on the ground directly by employing natural X-ray pulsar signal. In addition, it is expensive to perform flight demonstration. On the current statue of the study on X-ray pulsar-based navigation, it is better to establish a ground-based simulation and verification system to investigate and to verify the performance of each subsystem of X-ray pulsar-based navigation system.

7.1 Overall Design

7.1.1 Module Design

The scheme of ground-based simulation and verification system is shown in Fig. 7.1. This system is composed of the X-ray pulsar signal simulation module, the X-ray detection module, the pulsar signal processing module, the orbital dynamics simulation module, the navigation calculation module and the verification and evaluation module. Among them, the X-ray pulsar signal simulation module is designed for simulating the on-orbit observed X-ray pulsar signals generated at the spacecraft, the navigation detection module comprises the X-ray sensor and the atomic clock, the pulsar signal processing module processes the pulsar signals with the method provided in Chap. 3, the orbital dynamics simulation module provides a high-accuracy nominal orbital data according to the spacecraft dynamic orbit model provided in Chap. 2, inputs values for the simulation of X-ray pulsar on-orbit signals and references values for the verification and evaluation module, and the verification and evaluation module is designed for analyzing and evaluating the performance of the navigation system.

Among above modules, the X-ray pulsar signal simulation module and the X-ray detection module can be performed digitally or physically, while the orbital

W. Zheng and Y. Wang, *X-ray Pulsar-based Navigation*, Navigation: Science and Technology 5, https://doi.org/10.1007/978-981-15-3293-1_7

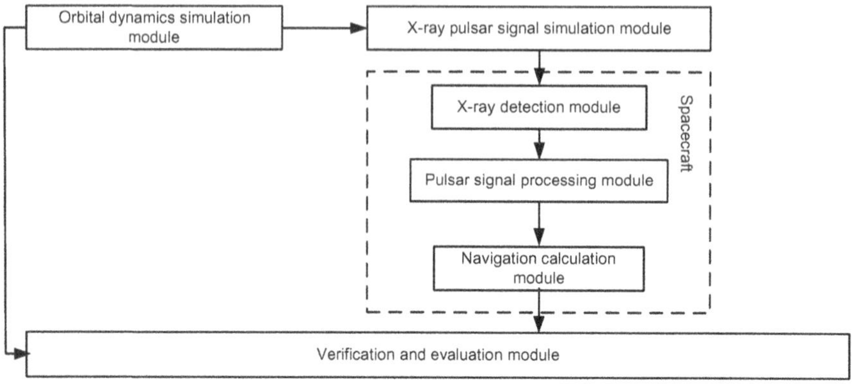

Fig. 7.1 Design drawing of ground-based simulation and verification system scheme

dynamics simulation module, the navigation calculation module and the verification and evaluation module can be performed digitally.

7.1.2 *Physics Configuration*

This simulation and verification system is composed of the spacecraft orbital dynamics simulation node, the X-ray pulsar signal simulation node, the simulation management node, the spacecraft navigation simulation node and the scene display node [1]. The Ethernet-based distributed simulation structure is adopted in the system, and the simulation management nodes are responsible for controlling the simulation progressing and simulation data recording of the entire system [2]. The simulation system structure is shown in Fig. 7.2 [3–8].

Among them,

- Simulation management node is designed for controlling the simulation flow, managing the simulation data, and performing the verification and evaluation;
- Spacecraft orbital dynamics simulation node is designed for simulating high-fidelity orbit of spacecraft;
- X-ray pulsar signal simulation node is designed for simulating the X-ray pulsar signal according to the modes including full-digital simulation, sensor response signal hardware simulation, and X-ray source and detector physical simulation;
- Spacecraft navigation simulation node is designed for processing pulsar signal and navigation calculation;
- Scene display node is designed for displaying the navigation process in a 3D simulation environment.

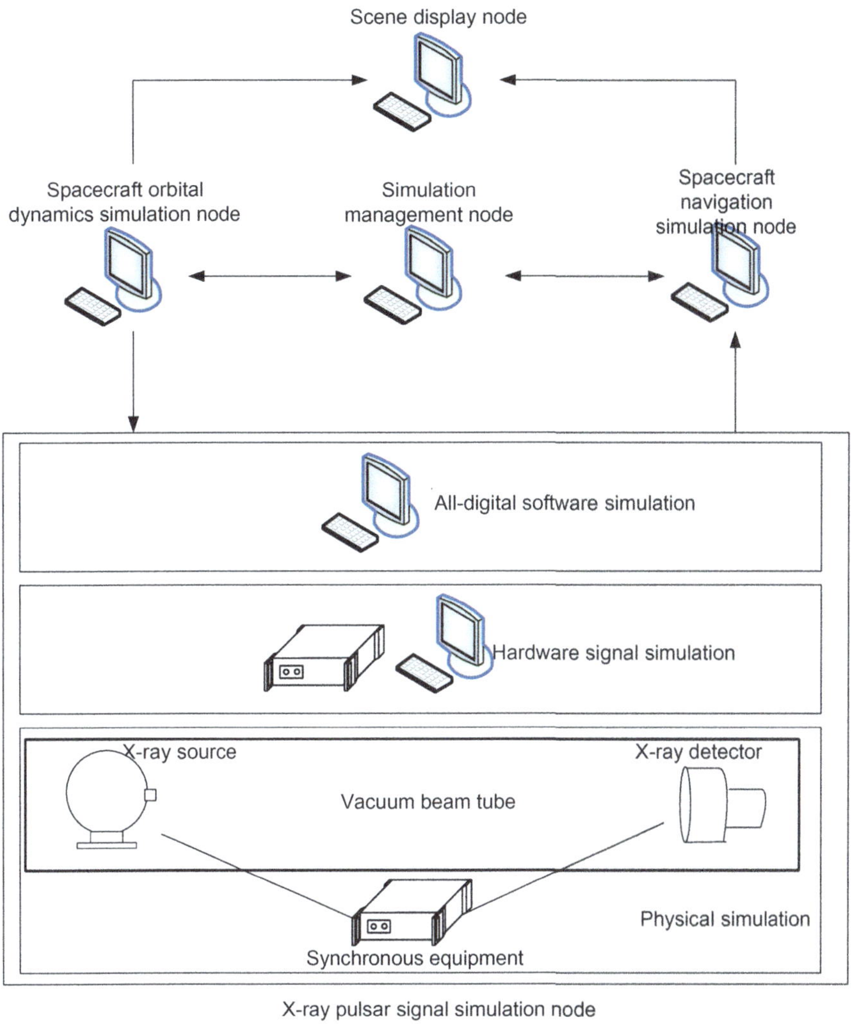

Fig. 7.2 Structure of ground-based simulation and verification system physics configuration

7.2 All-Digital Simulation and Verification Mode

7.2.1 A Design Framework of the Pulsar Signal Processing Software System

The object types, object parameters, task parameters, and task type setting in traditional pulsar signal processing software are fixed. Once users want to add new object types, object parameters and task parameters, they must rewrite the code of software.

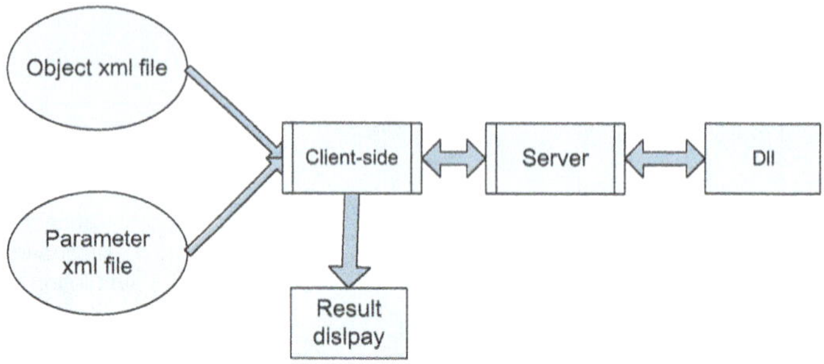

Fig. 7.3 Overall data flow chart

It causes the maintenance of software become expensive and the software be of little flexibility.

Thus, we developed a novel framework which aims to solve the above problems. The new framework uses independent dynamic libraries as the core and uses XML files as connection. In this way, users can add object types, object parameters and task parameters easily, and add tasks flexibly. The overall data flow chart and the task process chart are shown in Figs. 7.3 and 7.4.

7.2.2 *System Composition*

This system consists of object management module, task type management module and dynamic-link library management module.

7.2.2.1 Object Management Module

Object management module is used for managing object data which is related to the simulation mission. Users can add or delete object types in the object-type management table. There are several kinds of object types, such as pulsars, environment, spacecraft, sensors, radio telescope, etc. The object parameters can be entered by manually inputting or loading the XML file templates. The object management chart is shown in Fig. 7.5.

7.2.2.2 Task Type Management Module

Tasks of X-ray pulsar-based navigation are classified into several task modes. Every task belongs to a certain task mode and executes its procedure. When users choose one task mode, the system will call the corresponding procedure to handle the request of users. Currently, there are three modes at present, namely the common task mode,

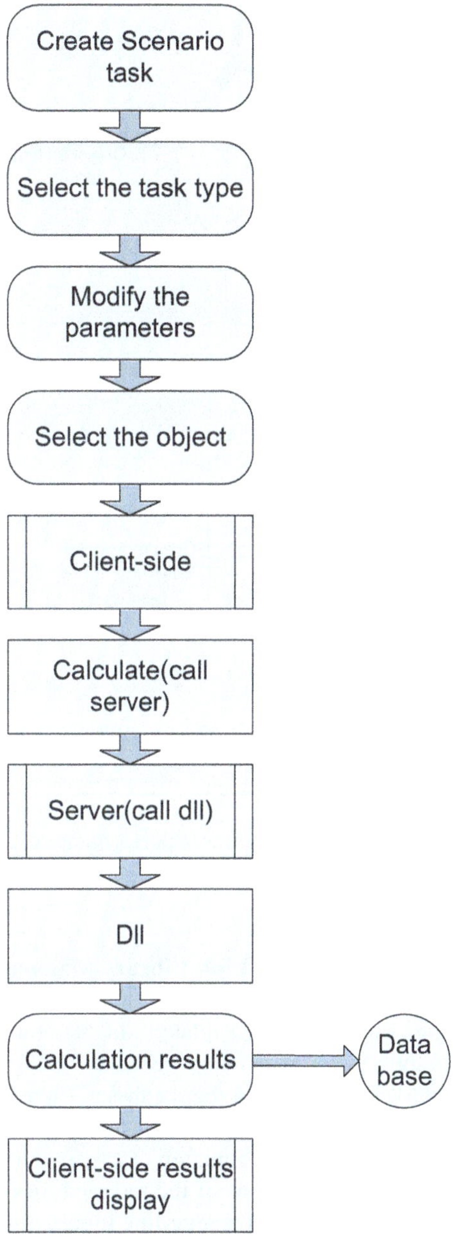

Fig. 7.4 Task process chart

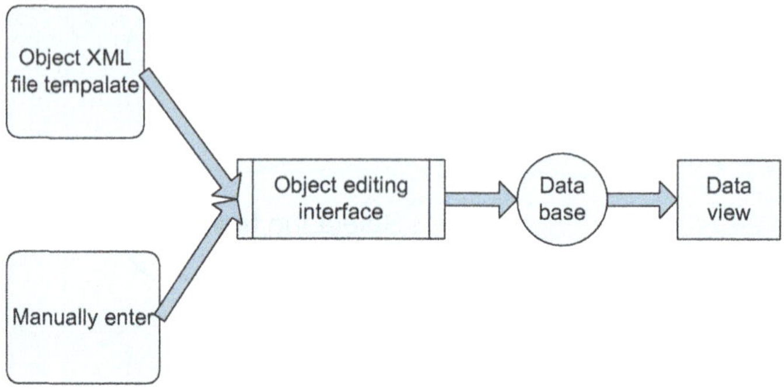

Fig. 7.5 Object management chart

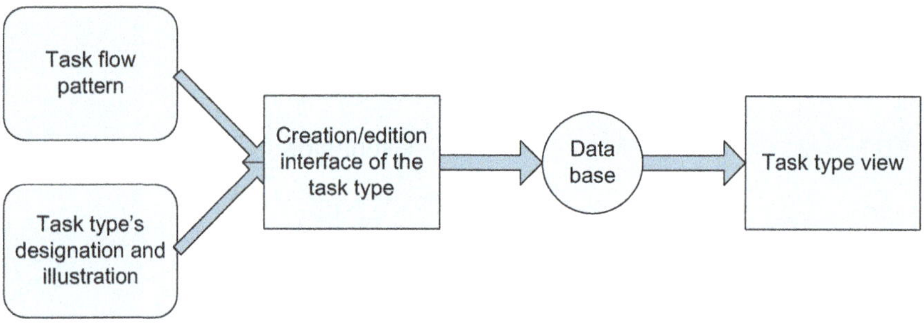

Fig. 7.6 Task management chart

data task mode and detector performance evaluation task mode. The task management chart is shown in Fig. 7.6.

7.2.2.3 Dynamic-Link Library Management Module

The system loads the related .dll files through dynamic-link library management module. When parameters necessary to the simulation are set already, these parameters will be saved in the database. There are mainly two types of parameters: (1) XML parameters which include the name of dynamic library, related task of dynamic library, the entrance function of dynamic library, the parameters of dynamic library, and etc.; and (2) Type of the returned files which indicate the way how the results are presented. The dynamic-link library management chart is shown in Fig. 7.7.

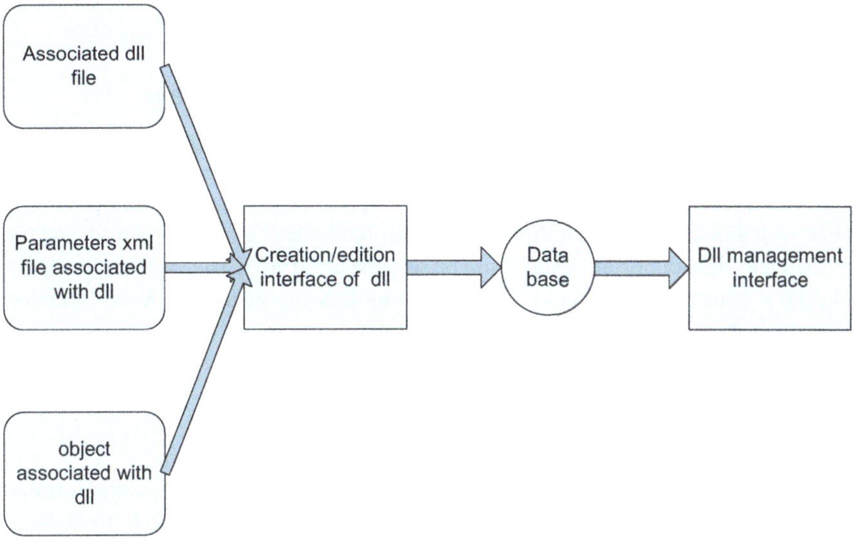

Fig. 7.7 Dynamic-link library management chart

7.2.3 Simulation Example

Take the pulsar static signal generation task as example to show how this system works. This task, used for generate the pulsar static initial data, belongs to the task mode of the signal analysis and simulation.

Step 1. Task creation
The users need to enter a task name in the interface of task creation.

Step 2. Parameter settings
The system will present task parameters, including group number, simulation duration, targeting steps, signal flux, noise flux, bin number, the bin for epoch folding, associated objects and so on.

Step 3. Running simulation
After the parameter settings, the user can click on the button of calculation, and then the background program will work and display the final results.

Step 4. Results display
After simulation, click the save icon in the menu bar to save the task. The result files will be written and saved at the location specified by users. And then, the result display interface will show the result.

From the discussion above, this software can fulfill the purpose that users can readily add and change the object types, object parameters and task parameters.

7.3 Semi-physical Simulation and Verification Mode

7.3.1 Components of Semi-physical Simulation System

In the semi-physical simulation system, the pulsar signal is produced and detected by physical devices instead of by digitally simulating. This system can provide a pulsar signal in the form of photon with a controllable flux, a variable period, and the inputted template of pulsar signal could be settable arbitrarily. The system consists of the central control unit, vacuum environment unit, X-ray generator, X-ray detector, software system and so on (Fig. 7.8).

7.3.1.1 X-Ray Generator

The X-ray generator is a gate-modulated tube. First of all, the pulsar profile is transformed to an appropriate voltage value which is used to drive the gate-modulated tube. If the filament's current and anode's voltage are fixed, the gate-modulated tube changes the output flux of X-ray by changing the voltage of the gate. So the flux of the X-ray is modulated by the time. As the X-ray tube has no window structure, the low energy X-ray is reserved (Fig. 7.9).

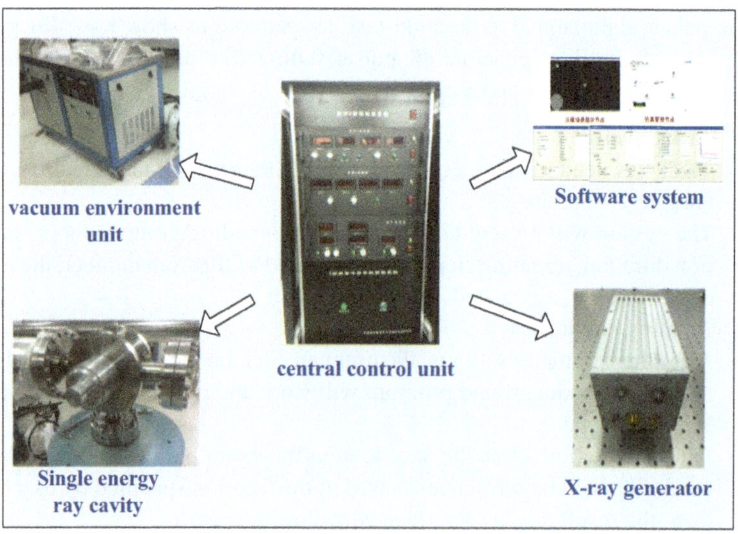

Fig. 7.8 Components of semi-physical simulation system

Fig. 7.9 The X-ray generator of the system

7.3.1.2 Vacuum Environment Simulation System

Regarding that the energy of most photons sent by an X-ray generator is below 2 keV, these photons can only be transmitted in a vacuum environment in that they will decay quickly in any other medium. This system uses high-performance compound molecular pump and rotary vane mechanical pump with a speed of 600 L/s to vacuumize the tube. It will take one hour to reach the vacuum degree of 10^{-4} Pa (Fig.7.10).

7.3.1.3 The Time Synchronization System for X-Ray Source and Detector

The traditional rotating-plate modulation scheme is difficult to implement the synchronization between X-ray source and detector. The system uses electronic device to modulate pulsar's signal, so it is much easier to trigger the X-ray source and detector synchronously. Furthermore, the atom clock is used to synchronize the interval of seconds (Fig. 7.11).

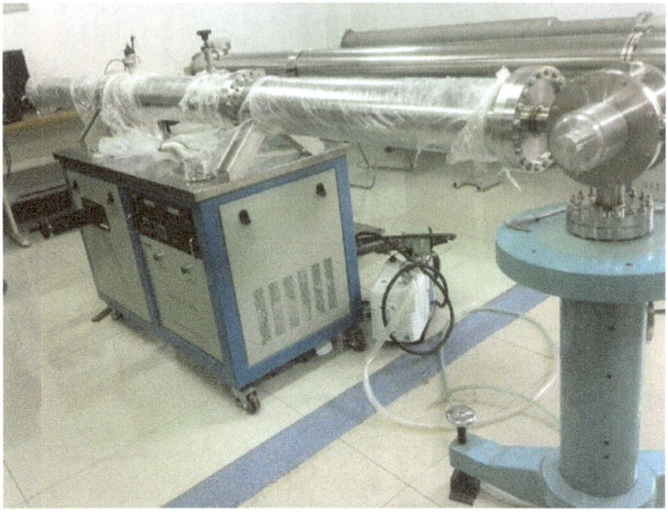

Fig. 7.10 Vacuum environment simulation system

Fig. 7.11 High-performance Rb atom clock

7.3.2 Dynamic Signal Simulation Experiment

When a spacecraft orbit, the signal received by the detector contains a Doppler frequency. To simulate this kind of signal requires a high-precision period tracking and flux control.

7.3.2.1 Experiments of Dynamic Signal for Low Earth Orbit

Select PSR B0531+21 as the observation object. Set the flux at 276.4793 ph/s. The simulation duration is 90 min. The initial state of spacecraft is [0 km, 6578 km, 0 km, −7784.3384 m/s, 0 m/s, 0 m/s]. Search the periods every 1 min and the period tracking results are shown in Fig. 7.12. The recovery profile is shown Fig. 7.13.

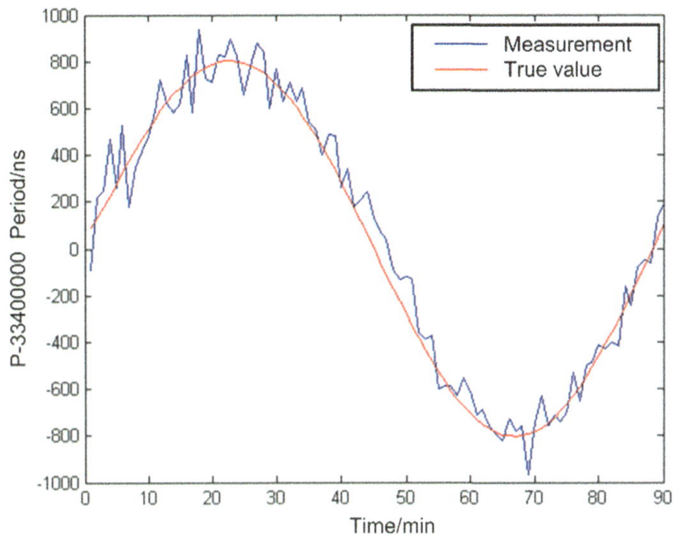

Fig. 7.12 The period tracking results

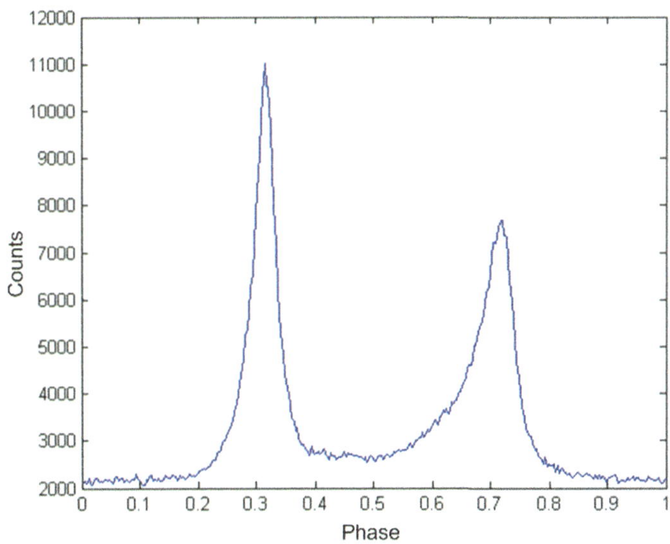

Fig. 7.13 The recovery profile using orbit information

7.3.2.2 Experiments of Dynamic Signal for High Earth Orbit

Select PSR B1937+21 as the observation object. Set the flux at 431.5924 ph/s. The simulation duration is 90 min. The orbit is low Earth orbit. The initial state of spacecraft is [−5654.543 km, 41617.618 km, 0 km, 1526.30 m/s, −207.37 m/s,

−2667.93 m/s]. Search the periods every 1 min and the period tracking results are shown in Fig. 7.14.

By employing the standard deviation as the criterion to evaluate the period tracking performance, the experiment results show that the precision of period tracking can reach the level of nanosecond.

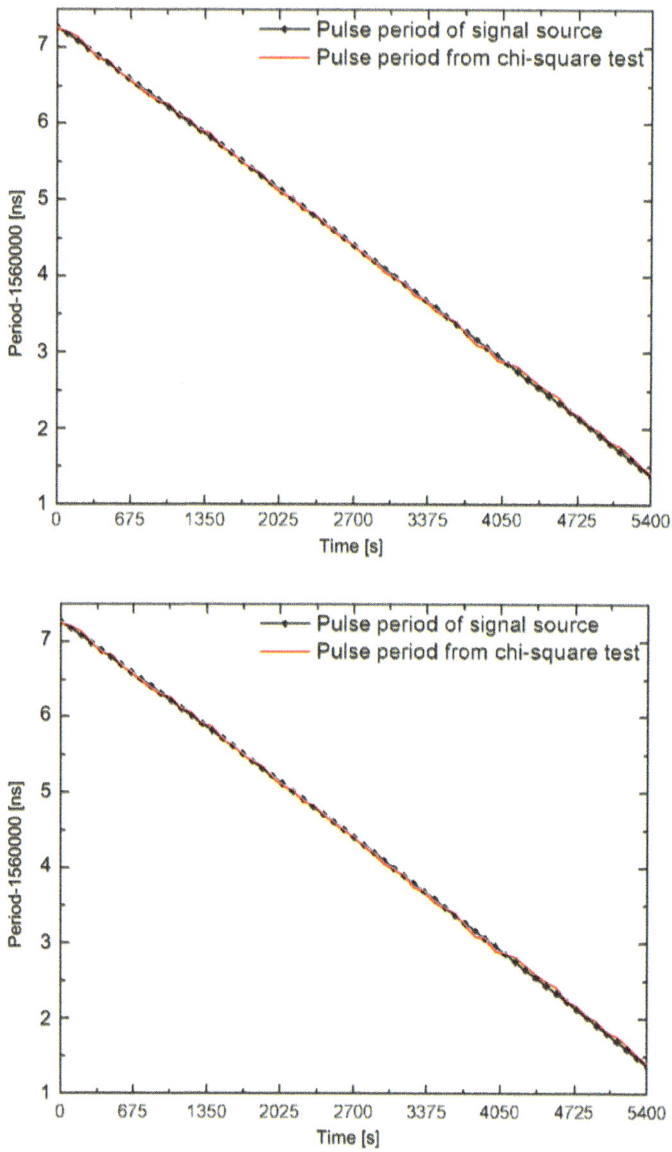

Fig. 7.14 The period tracking results

Table 7.1 Test results of single energy experiments

Character energy (keV)	FWHM (eV)
1.49	148
4.51	190
5.41	166
6.40	158
8.05	187
8.89	178
15.77	192

7.3.3 Energy Spectrum Experiment

7.3.3.1 Single Energy X-Ray Experiment

Single energy X-ray is used to calibrate the energy-efficiency curve of X-ray detectors. It is a significant parameter for a detector.

Set the anode high voltage of the X-ray source as 25 kV and the filament current as 2.2 A and the grid as 0 V. Use SDD Detector to collect photons. Get the energy spectrum, character energy (Kα) and FWHM by the data processing software (Table 7.1).

7.3.3.2 Hybrid Energy X-Ray Experiment

Hybrid energy X-ray is used to simulate the pulsar signal's energy spectrum. The hybrid energy spectrum can support the study on the energy spectrum of pulsars and test of the energy resolution of X-ray detector.

Set the anode high voltage of the X-ray source as 25 kV and the filament current as 2.2 A and the grid as 0 V. Use Mo target to generate the X-ray. The working duration of the system is one hour.

The hybrid energy spectrum of Mo target is shown in Fig. 7.15b. By comparing with the energy spectrum curve of PSR B0531+21, the hybrid energy spectrum is similar to it.

7.3.4 X-Ray Detector Test

7.3.4.1 SDD Detector Performance Test

1. Energy resolution and linear test

The test data of radioactive sources Fe^{55} and Am^{241} are used in this section to calibrate the relationship between detector photon energy and energy channel.

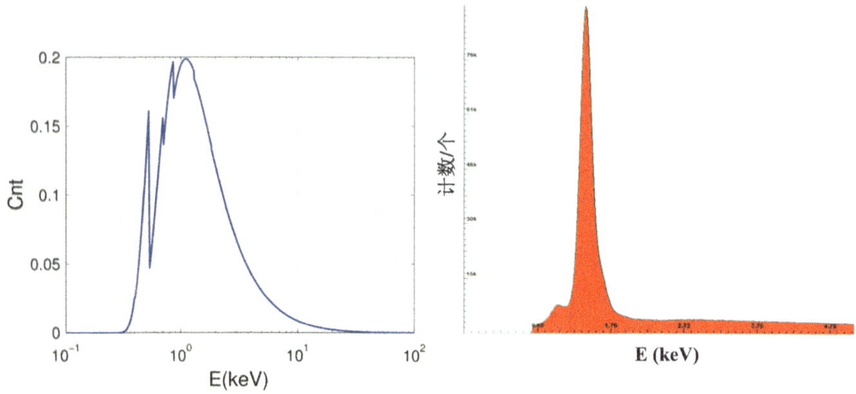

(a) Energy spectrum of PSR B0531+21 (b) The simulation energy spectrum

Fig. 7.15 Simulation of pulsar's energy spectrum

The energy spectrums of Fe^{55} and Am^{241} measured by the system are shown in Figs. 7.16 and 7.17, whose system energy resolutions (FWHM) are 155 eV@5.9 keV and 205 eV@13.93 keV respectively. Using the peak position and energy test data in Table 7.2, the fitted relationship between energy spectrum and number of energy channel is $E = 0.01178\ N + 0.06113$, where its linear correlation coefficient is 0.999

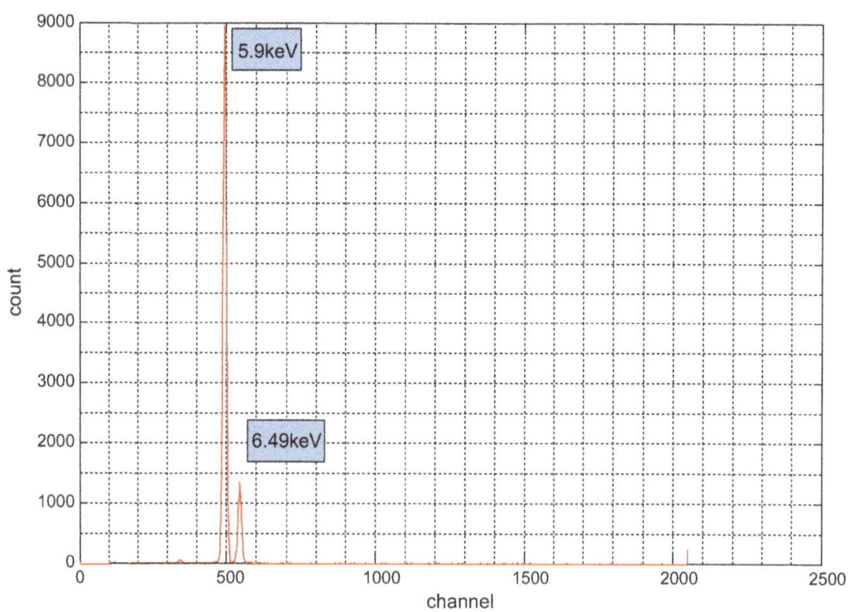

Fig. 7.16 Fe^{55} source (FWHM ~155 eV@5.9 keV)

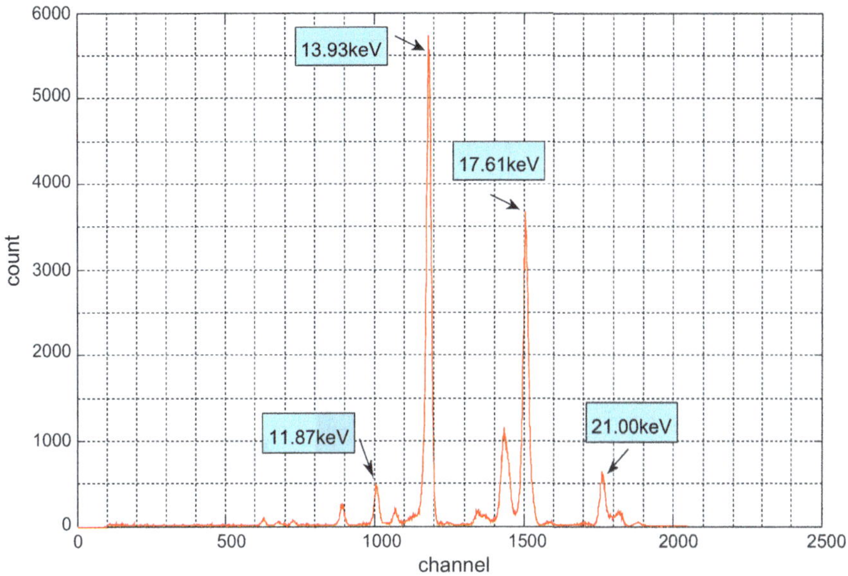

Fig. 7.17 Am^{241} source (FWHM ~205 eV@5.9 keV)

Table 7.2 Peak position and energy value

Source	Channel	Energy (keV)
Fe^{55}	492	5.90
	543	6.49
Am^{241}	1006	11.87
	1181	13.93
	1503	17.61
	1764	21.00

and its non-linearity is 1.05%. This demonstrates the energy of detector is almost linear to the number of energy channel.

2. System time accuracy and time resolution test

   a) Time accuracy test
      The time accuracy of detector is a crucial factor influencing the navigation accuracy during the overall process. A signal generator is used for generating the 5 kHz Gaussian pulses (Vpp = 1 V), and a data acquisition unit is used for measuring its counting accuracy. Based on the results shown in Fig. 7.18, most signals are recorded accurately and the mean error variance is 20.3 ns.
   b) Time resolution test
      The time resolution is a crucial index of detector, which refers to the minimum time interval between two photons received successively by a detector. The

Fig. 7.18 Time data recorded by data acquisition unit

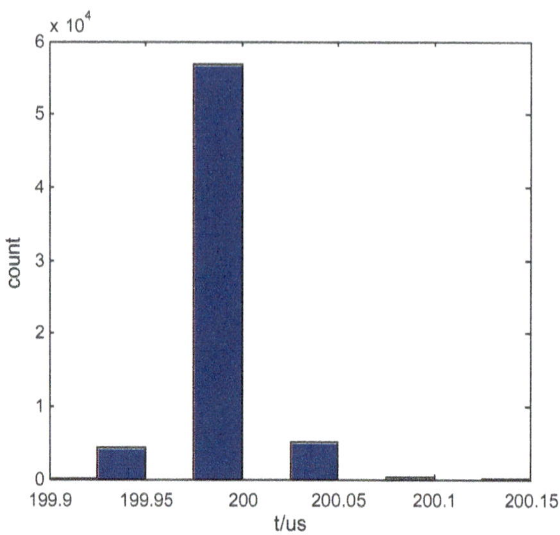

statistical distribution of interval of photon TOAs is shown in Fig. 7.19, and that of the interval of photon TOAs in the semi-log coordinate system (red crosses) is shown in Fig. 7.20 (the black point distribution is the theoretical value of exponential distribution). As shown by the two groups of data, the interval of photon TOAs is subject to an exponential distribution.

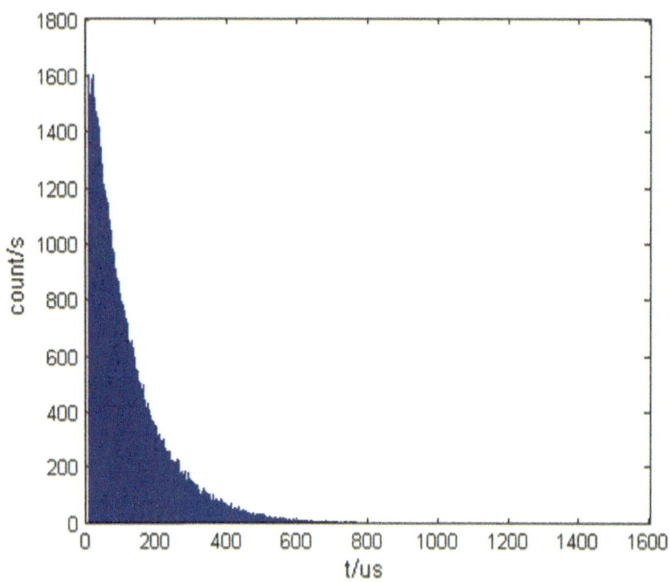

Fig. 7.19 Statistical distribution of interval of photon TOAs

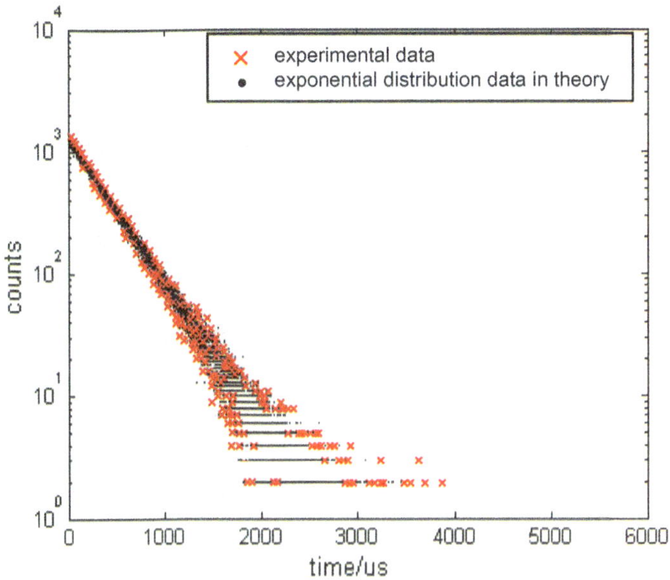

Fig. 7.20 Statistical distribution of interval of photon TOAs in the semi-log coordinate system

Amplifying the front part of Fig. 7.19, we obtain Fig. 7.21. As can be seen from Fig. 7.21, the minimum time interval between two neighboring photons is about 10 μs, and the test results in other conditions is the same as this one. Therefore, the time resolution of this detector can be viewed as 10 μs.

7.3.4.2 SCD Detector Performance Test

Because of the principle of the SCD detector, the response of SCD detector has a time delay which must be calibrated in order to ensure a satisfactory series of photon TOAs. The schematic setup of the experiment is shown in Fig. 7.22.

The profile of PSR B0531+21 is used in this experiment, and SCD and SDD Detector keep collecting photon data for one hour. Figure 7.23 shows two profiles obtained by SDD and SCD data respectively. By comparing of the two profiles, the time delay of SCD detector can be determined as 11.3525 ms.

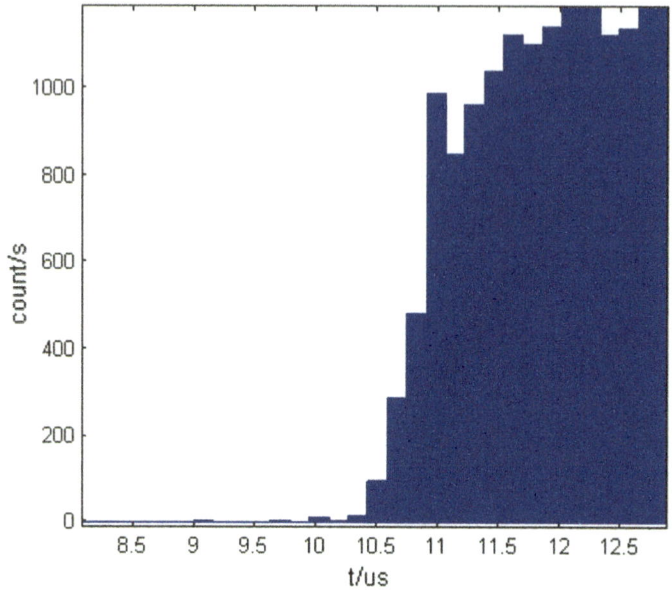

Fig. 7.21 Time resolution of system

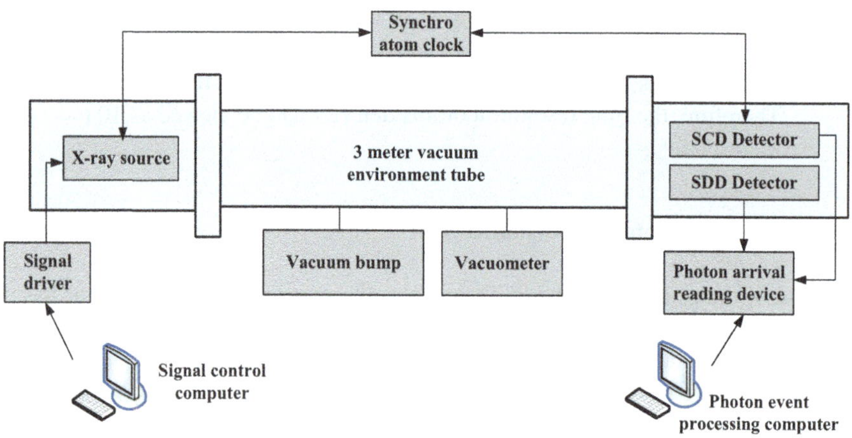

Fig. 7.22 Setup of the time delay calibration testbed

7.4 Summary

The ground-based simulation and verification system for X-ray pulsar-based navigation is designed and initially realized in this chapter. The system is composed of the X-ray pulsar signal simulation module, the X-ray detection module, the pulsar signal processing module, the orbital dynamics simulation module, the navigation

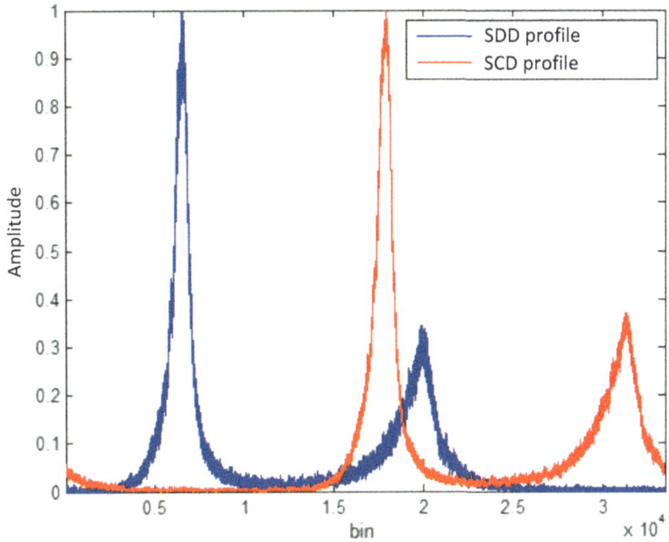

Fig. 7.23 Time delay of the SCD detector

calculation module and the verification and evaluation module. It has two simulation and verification modes, namely, full-digital mode and semi-physical mode. In the full-digital mode, the numeric simulation technique is adopted in the system to simulate the whole process from generation of X-ray pulsar photons to navigation calculation. In the semi-physical mode, a high-accuracy X-ray periodic signal simulator is established in the system to simulate the generation of X-ray pulsar signals and an X-ray detector is used for detection of X-ray photons. The performance of the ground-based simulation and verification system is verified by dynamic signal simulation experiment, energy spectrum experiment, X-ray detector tests.

References

1. Sun S (2011) Research on spacecraft autonomous navigation method based on X-ray pulsars. National University of Defense Technology, Changsha
2. Wang Y (2011) Research on X-ray pulsar-based navigation method in deep space exploration. National University of Defense Technology, Changsha
3. Zhou J (2008) Research on distributed simulation system design and application. National University of Defense Technology, Changsha
4. Liu L, Zheng W, Tang G (2012) X-ray pulsar-based navigation semi-physical simulation system. J Natl Univ Def Technol 34(5):10–14
5. Zhen W, Tang G, Sun S, Chang S. Periodic ray signal generating system. Chinese patent: ZL200810031475.1
6. Zheng W, Sun S, Tang G, Liu L. X-ray pulsar-based navigation semi-physical simulation system. Chinese patent: ZL 201010022035

7. Liu L, Yu H, Zheng W (2014) Measurement of X-ray photon energy and arrival time using a silicon drift detector. Chin Phys C
8. Liu L (2014) Navigation constellation orientation parameter determination based on X-ray pulsars. National University of Defense Technology, Changsha

The manufacturer's authorised representative in the EU is Springer Nature Customer Service Centre GmbH, Europaplatz 3, 69115 Heidelberg, Germany. If you have any concerns regarding our products, please contact ProductSafety@springernature.com

Printed and bound by CPI Group (UK) Ltd, Croydon, CR0 4YY
30/07/2026
02181693-0005